Meikang Qiu and Edwin Sha

Heterogeneous Parallel Embedded Systems

Meikang Qiu and Edwin Sha

Heterogeneous Parallel Embedded Systems

Time and Power Optimization

VDM Verlag Dr. Müller

Impressum/Imprint (nur für Deutschland/ only for Germany)
Bibliografische Information der Deutschen Nationalbibliothek: Die Deutsche Nationalbibliothek
verzeichnet diese Publikation in der Deutschen Nationalbibliografie; detaillierte bibliografische
Daten sind im Internet über http://dnb.d-nb.de abrufbar.
Alle in diesem Buch genannten Marken und Produktnamen unterliegen warenzeichen-, marken-
oder patentrechtlichem Schutz bzw. sind Warenzeichen oder eingetragene Warenzeichen der
jeweiligen Inhaber. Die Wiedergabe von Marken, Produktnamen, Gebrauchsnamen,
Handelsnamen, Warenbezeichnungen u.s.w. in diesem Werk berechtigt auch ohne besondere
Kennzeichnung nicht zu der Annahme, dass solche Namen im Sinne der Warenzeichen- und
Markenschutzgesetzgebung als frei zu betrachten wären und daher von jedermann benutzt
werden dürften.

Coverbild: www.purestockx.com

Verlag: VDM Verlag Dr. Müller Aktiengesellschaft & Co. KG
Dudweiler Landstr. 125 a, 66123 Saarbrücken, Deutschland
Telefon +49 681 9100-698, Telefax +49 681 9100-988, Email: info@vdm-verlag.de
Zugl.: Dallas, University of Texas at Dallas, Diss., 2007

Herstellung in Deutschland:
Schaltungsdienst Lange o.H.G., Zehrensdorfer Str. 11, D-12277 Berlin
Books on Demand GmbH, Gutenbergring 53, D-22848 Norderstedt
Reha GmbH, Dudweiler Landstr. 99, D- 66123 Saarbrücken
ISBN: 978-3-639-09619-4

Imprint (only for USA, GB)
Bibliographic information published by the Deutsche Nationalbibliothek: The Deutsche
Nationalbibliothek lists this publication in the Deutsche Nationalbibliografie; detailed
bibliographic data are available in the Internet at http://dnb.d-nb.de.
Any brand names and product names mentioned in this book are subject to trademark, brand or
patent protection and are trademarks or registered trademarks of their respective holders. The use
of brand names, product names, common names, trade names, product descriptions etc. even
without
a particular marking in this works is in no way to be construed to mean that such names may be
regarded as unrestricted in respect of trademark and brand protection legislation and could thus
be used by anyone.

Cover image: www.purestockx.com

Publisher:
VDM Verlag Dr. Müller Aktiengesellschaft & Co. KG
Dudweiler Landstr. 125 a, 66123 Saarbrücken, Germany
Phone +49 681 9100-698, Fax +49 681 9100-988, Email: info@vdm-verlag.de

Copyright © 2008 VDM Verlag Dr. Müller Aktiengesellschaft & Co. KG and licensors
All rights reserved. Saarbrücken 2008

Produced in USA and UK by:
Lightning Source Inc., 1246 Heil Quaker Blvd., La Vergne, TN 37086, USA
Lightning Source UK Ltd., Chapter House, Pitfield, Kiln Farm, Milton Keynes, MK11 3LW, GB
BookSurge, 7290 B. Investment Drive, North Charleston, SC 29418, USA
ISBN: 978-3-639-09619-4

To my dearest family

ACKNOWLEDGEMENTS

First, I would like to thank my adviser, Dr. Edwin Sha, for his ideas, directions, and constant encouragement. Without his help and support, this body of work would not have been possible. I also thank committee member Dr. Farokh B. Bastani, Dr. Kang Zhang, and Dr. Ying Liu for their service and help in improving this document.

Next, I thank the other members of Dr. Sha's research group: Nelson Passos, Sissades Tongsima, Qingfeng Zhuge, Zhong Wang, and Zili Shao, for their contributions to certain aspects of my research. I also would like to thank Chun Xue, Meilin Liu, Kevin Chen, and Ying-Ju Suen for making the lab always full of laughter.

I also thank all my teachers from whom I learned so much in my long journey of formal education. Specially, in Shanghai Jiao Tong University, they are Prof. Shugang Yang, Prof. Jiliang Xu, and Prof. Jingtao Shi. In Erik Jonsson School of Engineering and Computer Science of University of Texas at Dallas, they are Dr. S. Q. Zheng, Dr. Ding-Zhu Du, Dr. Hal Sudborough, Dr. Gopal Gupta, Dr. Weili Wu, Dr. G. R. Dattatreya, Dr. Poras T. Balsara, and many others.

Furthermore, I would like to thank our department head, Dr. D. T. Huynh, for his aid to me during the completion of my program. Also, I thank the university and TI's University Program for their financial support of my studies.

I also appreciate the help from the secretaries and staff in the Department of Computer Science: Judy Patterson, Emebet Sahle, Rosa Diaz, Julie Weekly, Lynda Gary, Norma Richardson, and Brain Nelson, for their supports.

Finally, but most significantly, I thank my wife Diqiu Cao and my son David Qiu, my father Shiqing Qiu and mother Longzhi Yuan, my brother Meisheng Qiu and sister Meitang Qiu, and many other relatives, for their continuous love, support, trust, and encouragement through the whole trip of my life. Without them, none of this would have happened.

February 2007

TABLE OF CONTENTS

CHAPTER 1

INTRODUCTION

Embedded systems are driving an information revolution with their pervasion in our everyday lives. These tiny, quick and smart systems can be found everywhere, ranging from consumer electronics such as cell phones, cameras, refrigerators, TVs, to critical infrastructure such as telecommunication networks, electrical power grids, financial institutions, nuclear plants. With more and more embedded systems networked, embedded systems and their applications are radically changing the way that the information is collected, shared and processed. For example, networks with thousands or millions of embedded sensor systems could monitor the environment, the battlefield, or the factory; embedded microscopic sensor systems could traverse the bloodstream, monitor health conditions and report them wirelessly; intelligent devices with smart embedded systems could be incorporated into the wings of aircraft to allow fine-grained control of airflow.

The increasingly ubiquitous embedded systems pose a host of technical challenges different from those faced by general-purpose computers because they are more application specific and more constrained in terms of timing, power, area, memory and other resources. It becomes an important research problem to design high-performance and low-power embedded systems with various constraints and limited resources. In this dissertation, we have attacked this problem from various aspects including high-level architecture synthesis, low power optimization, and energy and performance optimization for memory in embedded system research.

1

High-level architecture synthesis is a common and critical step in the designs of high-performance, low power and reliable embedded systems. Given various constraints and limited resources, high-level architecture synthesis can produce an application-specific architecture with optimization of timing, power, reliability and other parameters. In this dissertation, we study the problem of architecture synthesis for real-time DSP, one of the most popular embedded applications. DSP applications that process signals by digital means need special high-speed functional units (FUs) like adders and multipliers to perform addition and multiplication operations. With more and more different types of FUs available, same type of operations can be processed by heterogeneous FUs with different costs, where the cost may relate to power, reliability, etc. Furthermore, some tasks may not have fixed execution time. Such tasks usually contain conditional instructions and/or operations that could have different execution times for different inputs. Therefore, for such special purpose architecture synthesis, an important problem is how to assign a proper FU type to each operation of a DSP application and generate a schedule in such a way that we can minimize the total costs while satisfying timing constraints with guaranteed confidence probabilities. We proposed several efficient algorithms to solve the problem. The experiments show that our algorithms can effectively reduce the total cost compared with the previous work.

In low power design, our focus is to optimize power consumption in embedded systems. Low power is becoming a critical design issue and performance metric in embedded system design due to wide use of portable devices, especially those powered by batteries. DSP processor has multiple FUs and can process several instructions simultaneously. While this multiple-FU architecture can be exploited to increase instruction-level parallelism and improve time performance, it causes more power consumption. In embedded systems, high performance DSP needs to be performed not only with high data throughput but also with low power consumption. Therefore, it becomes an important problem to reduce the power consumption of a DSP. To solve this problem, several techniques have been proposed. We combined *Dynamic Voltage Scaling* (DVS) and soft real-time to solve the *Voltage Assignment with Probability* (VAP) Problem. VAP problem involves finding a voltage level to be used for each node of an *Probabilistic Date Flow Graph* (PDFG) in uniprocessor and multiprocessor DSP systems. This work tremendously improves the state-of-the-art techniques on power reduction for real-time embedded systems.

Another application is heterogeneous sensor network. We applied our efficient algorithms to dynamic adjust the working mode of sensors and achieved significant energy saving. As we know, energy and timing are critical issues for wireless sensor networks since most sensors are equipped with non-rechargeable batteries that have limited lifetime. However, sensor nodes usually work under dynamic changing and hard-to-predict environments. We use a novel *adaptive online energy-saving* (AOES) algorithm to save total energy consumption for heterogeneous sensor networks. We propose an optimal sub-algorithm *MAP_Opt* to minimize the total energy consumption while satisfying the timing constraint with a guaranteed confidence probability.

Also, we design new rotation scheduling algorithms for real-time applications that produce schedules consuming minimal energy. In our algorithms, we use rotation scheduling to get schedules for loop applications. The schedule length will be reduced after rotation. Then, we use DVS to assign voltages to computations individually in order to decrease the voltages of processors as much as possible within the timing constraint. The experimental results show that this approach can further reduce the total energy consumption. On average, our algorithm shows a 32.6% reduction in energy consumption compared with the ILP technique in [110] for hard real-time and even better for soft real-time.

Furthermore, we combine data mining and prefetching to reduce energy consumptions. First, we use data mining to predict the distribution of execution time and find the association rules between execution time and different inputs from history table. Then we use rotation scheduling to obtain the best assignment for total cost minimization. Finally, we use prefetching to prepare data in advance at run time. Experiments demonstrate the effectiveness of our algorithm. Our approach can handle loops efficiently. In addition, it is suitable to both soft and hard real-time systems.

Many high-performance DSP processors employ multi-bank on-chip memory to improve performance and energy consumption. This architectural feature supports higher memory bandwidth by allowing multiple data memory accesses to be executed in parallel. However, making effective use of multi-bank memory remains difficult, considering the combined effect of performance and energy requirement. In this project, our focus is to study the scheduling and assignment problem that minimizes the total energy while satisfying performance requirements. Our approach has several major contributions: First, we study the combined effects of energy-saving and performance of memory and ALU in a systematic approach. Second, we exploit the energy saving of memory with operation scheduling and memory type assignment. Third, data locality has been improved by using variable partition.

In this dissertation, we have focused on developing models, methodologies, and algorithms for high-performance, low-power embedded systems. In the following, a brief introduction for our research is given. The rest of this dissertation is organized as follows: In Section 1.1, the related work is presented. Section 1.2 presents the overview of our techniques. The contributions of this dissertation is summarized in Section 1.3. Section 1.4 presents the outline of the dissertation.

1.1 Related Work

There have been extensive studies in the fields of high-level architecture synthesis, low-power optimization, and energy & performance optimization for memory in embedded system research. The relationships between our research results and the prior results are summarized as follows.

1.1.1 High-Level Architecture Synthesis

Our high-level architecture synthesis research focuses on developing effective architecture synthesis techniques for real-time DSP (Digital Signal Processing) with heterogeneous functional units. There have been a lot of research efforts in this field. The road map shown in Figure 1.1 illustrates the relationships between our research results and the prior results in high-level architecture synthesis for DSP systems. As shown in Figure 1.1, There have been a lot of research efforts on allocating applications in heterogeneous distributed systems [4–8,10,33,77]. Incorporating reliability cost into heterogeneous distributed systems, the reliability driven assignment problem has been studied in [26,81,86]. In these work, allocation are performed based on a fixed architecture. However, when performing assignment in architecture synthesis, no fixed architectures are available. Most previous work on the synthesis of special purpose architectures for real-time DSP applications focuses on the architectures that only use homogeneous FUs, that is, same type of operations will be processed by same type of FUs [14,17,29,39,42,52,56,57,62,65].

Our work of Chapter 2 is related to the work in [43,80,92]. *Probabilistic retiming* (PR) had been proposed in [64,92]. For a system without resource constraints, PR can be applied to optimize the input graph, i.e., reduce the length of the longest path of the graph such that the probability of the longest path computation time being less than or equal to the given timing constraint, L, is greater than or equal to a given confidence probability P.

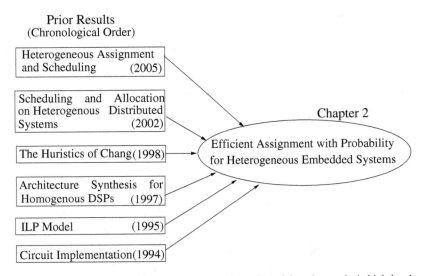

Figure 1.1. The relationships between our research results and the prior results in high-level architecture synthesis for DSP systems.

Since the execution times of the nodes can be either fixed or varied, a probability model is employed to represent the execution time of the tasks. But PR does not model the hard HA problem which focuses on how to obtain the best assignment from different FU types. In [80], the authors proposed two optimal algorithms for the hard HA problem when the given input is a tree or simple path, and three heuristic algorithms for the general hard HA problem. But they do not consider varied execution time situation. Also, for the general problem, their solutions are not optimal. Although the ILP model from [43] can obtain an optimal solution for the heterogeneous assignment problem, it is a NP-hard problem to solve the ILP model. Therefore, the ILP model may take a very long time to get results even when a given *Data Flow Graph* (DFG) is not very big. In Chapter 2, the algorithm solving the hard HA problem in [43] is called *hard HA ILP algorithm* [43] in general.

1.1.2 Low Power Design

Our low power optimization research focuses on reducing power consumption on various embedded systems by developing effective and efficient type (voltage) assignment and scheduling techniques. The road map shown in Figure 1.2 illustrates the relationships between our research results and the prior results in low power design.

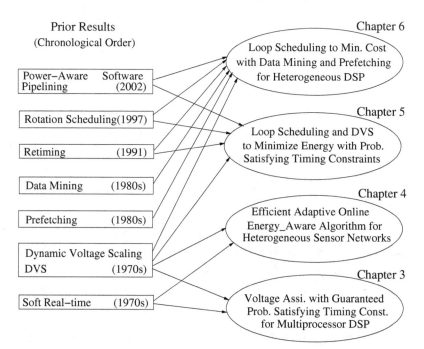

Figure 1.2. The relationships between our research results and the prior results in low power design.

As shown in Figure 1.2, we have done several work on low power design for embedded systems, such as DVS and loop scheduling to minimize energy consumption for DSP. The increasingly ubiquitous DSP systems pose great challenges different from those faced by general-purpose computers. DSP systems are more application specific and more constrained in terms of power, timing, and other resources. Energy-saving is a critical issue and performance metric in DSP systems design due to wide use of portable devices, especially those powered by batteries [16, 17, 28, 50, 65, 101]. The systems become more and more complicate and some tasks may not have fixed execution time. Such tasks usually contain conditional instructions and/or operations that could have different execution times for different inputs [36–38, 92, 111]. It is possible to obtain the execution time distribution for each task by sampling and knowing detailed timing information about the system or by profiling the target hardware [91]. Also some multimedia applications, such as image, audio, and video data streams, often tolerate occasional deadline misses without being noticed by human visual and auditory systems. For example, in packet audio applications, loss rates between 1% - 10% can be tolerated [11].

Prior design space exploration methods for hardware/software codesign of DSP systems [42, 43, 80, 95] guarantee no deadline missing by considering worst-case execution time of each task. Many design methods have been developed based on worst-case execution time to meet the timing constraints without any deadline misses. These methods are pessimistic and are suitable for developing systems in a hard real-time environment, where any deadline miss will be catastrophic. However, there are also many soft real-time systems, such as heterogeneous systems, which can tolerate occasional violations of timing constraints. The above pessimistic design methods can't take advantage of this feature and will often lead to over-designed systems that deliver higher performance than necessary at the cost of expensive hardware, higher energy consumption, and other system resources.

There are several papers on the probabilistic timing performance estimation for soft real-time systems design [36–38, 45, 68, 91, 92, 111]. The general assumption is that each task's execution time can be described by a discrete probability density function that can be obtained by applying path analysis and system utilization analysis techniques. Hu et al. [111] propose a state-based probability metric to evaluate the overall probabilistic timing performance of the entire task set. However, their evaluation method becomes very time consuming when task has many different execution time variations. Hua et al. [37, 38] propose the concept of *probabilistic design* where they design the system to meet the timing constraints of periodic applications statistically. But their algorithm is not optimal and only suitable to uniprocessor executing tasks according to a fixed order, that is, a simple path. In Chapter 3, we will propose an optimal algorithm for the uniprocessor situation. Also, we will give an optimal algorithm for multiprocessor executing tasks according to an executing order in DAG (Direct Acyclic Graph).

Low power and low energy consumptions are extremely important for real-time DSP systems. Dynamic voltage scaling (DVS) is one of the most effective techniques to reduce energy consumption [20, 79, 82, 110]. In many microprocessor systems, the supply voltage can be changed by mode-set instructions according to the workload at run-time. With the trend of multiple cores being widely used in DSP systems, it is important to study DVS techniques for multiprocessor DSP systems. Chapter 3 focuses on minimizing expected energy consumption with guaranteed probability satisfying timing constraints via DVS for real-time multiprocessor DSP systems.

In Chapter 4, we study the energy saving issue in heterogeneous sensor networks. As we know, energy and timing are critical issues for wireless sensor networks since most sensors are equipped with non-rechargeable batteries that have limited lifetime. However, sensor nodes usually work under dynamic changing and hard-to-predict environments. This chapter uses a novel *adaptive online energy-saving* (AOES) algorithm to save total energy consumption for heterogeneous sensor networks. Due to the uncertainties in execution time of some tasks and multiple working mode of each node, this paper models each varied execution time as a probabilistic random variable to save energy by selecting the best mode assignment for each node, which is called MAP (Mode Assignment with Probability) problem in this paper. We propose an optimal sub-algorithm *MAP_Opt* to minimize the total energy consumption while satisfying the timing constraint with a guaranteed confidence probability.

In Chapter 5, we use loop scheduling to further extend our work in Chapter 3. Low energy consumption is an important problem in real-time embedded systems and loop is the most energy consuming part in most cases. In this chapter, we use the same model as that in Chapter 3, i.e., we model each varied execution time as a probabilistic random variable. To further optimize our algorithm, we use rotation scheduling and DVS (Dynamic Voltage Scaling) to minimize the expected total energy consumption while satisfying the timing constraint with a guaranteed confidence probability. Our approach can handle loops efficiently. In addition, it is suitable to both soft and hard real-time systems. And even for hard real-time, we have good results. The experimental results show that our approach achieves significant energy saving than list scheduling and ILP (Integer Linear Programming) voltage assignment.

We combine loop scheduling and data mining and prefetching to solve the heterogeneous FU assignment problem, which is shown in Chapter 6. We propose an algorithm, *LSHAP*, to give the efficient solutions. We first use data mining to obtain the PDF of execution times, which are modeled as random variable in this paper. And find the association rules between execution times and different inputs. Then, by taking advantage of the uncertainties in execution time of tasks, we give out FU assignments and scheduling to minimize the expected total cost while satisfying timing constraint with guaranteed probabilities. We repeatedly regroup a loop based on rotation scheduling and decrease the energy by voltage selection as much as possible within a timing constraint. Finally, we prefetch the data needed in advance at run time. Our approach can handle loops efficiently. Experimental results show the significant cost-saving of our approach.

1.1.3 Energy and Performance Optimization for Memory

In many advanced memory architecture, there are heterogeneous memory banks. Different memory banks have different memory access time latency and energy consumption for same operations [4, 26, 42, 80]. A certain memory type may access the data stored slower but with less energy consumption, while another type will access the data faster with higher energy consumption. Also, there is a limitation of how many banks can be accessed simultaneously in certain memory architectures. Therefore, an important problem arises: how to partition variable to different banks and assign a type to each bank for an application such that the timing constraint can be satisfied and the total energy consumption can be minimized.

As shown in Figure 1.3, much research has been conducted in the area of using multi-bank memory to achieve maximum instruction level parallelism, i.e., optimize performance [21, 51, 54, 78, 88, 103]. These approaches differ in either the models or the heuristics. However, these works seldom consider the combined effect of performance and energy requirements. Actually, performance requirement often conflicts with energy saving [22,23,48,63,94,106]. Hence, significant energy saving and performance improvements can be obtained by exploiting heterogeneous multi-bank memory at the instruction level. Wang et al. [97] have considered the combined effect. They proposed the VPIS algorithm to overcome it. But their algorithm does not fully exploit the heterogeneous multi-bank memory architecture in energy saving.

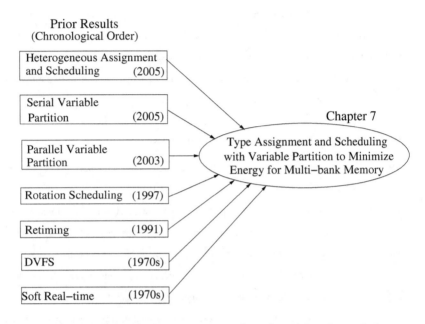

Figure 1.3. The relationships between our research results and the prior results in energy and performance optimization for memory.

Combining the consideration of energy and performance, in Chapter 7, we propose a novel memory model to overcome the weakness of previous techniques. An algorithm, TAMRS (*Type Assignment and Minimum Resource Scheduling*), is proposed. The algorithm use type assignment with the consideration of variable partition to find the best configuration for both memory and ALU to minimize total energy while satisfying performance requirements.

1.2   Technical Overview

In this section, we provide a technical overview on our research work in this dissertation. In the following, the overview for our heterogeneous assignment to minimize cost while satisfying hard/soft timing constraints is presented in Section 1.2.1, and the overview for our voltage assignment with guaranteed probability satisfying timing constraint for real-time multiproceesor DSP is shown in Section 1.2.2. Our efficient adaptive online energy-aware algorithm for heterogeneous sensor networks is summarized in Section 1.2.3. The energy minimization with guaranteed probability satisfying timing constraint via DVS and loop scheduling is summarized in Section 1.2.4. Our novel approach using data mining and prefetching to minimize cost for heterogeneous DSP is shown in Section 1.2.5. Finally, we summarize our heterogeneous scheduling and assignment with variable partition to minimize energy for multi-bank memory in Section 1.2.6.

1.2.1   Heterogeneous Assignment to Minimize Cost while Satisfying Hard/Soft Timing Constraints

It is a critical issue to do high-level synthesis for special purpose architectures of real-time embedded systems to satisfy the time and cost requirements [16, 17, 28, 42, 43, 50, 65, 95, 101]. The cost of embedded systems may relate to power, reliability, etc. [26, 32, 81, 86]. The systems become more and more complicate in two aspects. First, in many systems, such as heterogeneous parallel DSP systems [4, 33], same type of operations can be processed by heterogeneous FUs with different costs. Therefore, an important problem arises: how to assign a proper FU type to each operation of a DSP application such that the requirements can be met and the total cost can be minimized while satisfying timing constraints with a guaranteed confidence probability [80].

Second, some tasks may not have fixed execution time. Such tasks usually contain conditional instructions and/or operations that could have different execution times for different inputs [36–38,92,111]. Although many static assignment techniques can thoroughly check for the best assignment for dependent tasks, existing methods are not able to deal with such uncertainty. Therefore, either worst-case or average-case computation times for these tasks are usually assumed. Such assumptions, however, may not be applicable for real-time systems and may result in an uneffective task assignment. Using probabilistic approach, we can obtain solutions that can not only be used for hard real-time systems, but also provide more choices of smaller total costs while satisfying timing constraints with guaranteed confidence probabilities.

This dissertation presents high-level synthesis algorithms which operate in proba-bilistic environments to solve the *heterogeneous assignment with probability* (HAP) prob-lem. In the HAP problem, we model the execution time of a task as a random variable [108]. For heterogeneous systems, each FU type has different cost, representing hardware cost, size, reliability, etc. Faster one has higher cost while slower one has lower cost. Chapter 2 shows how to assign a proper FU type to each node of a *Probabilistic Data Flow Graph* (PDFG) such that the total cost is minimized while satisfying the timing constraint with a guaranteed confidence probability. In other words, we can guarantee that the total exe-cution time of the PDFG is less than or equal to the timing constraint with a probability greater than or equal to P.

It is known that the hard HA problem is NP-complete [80]. Since the HAP problem is NP harder than the hard HA problem, the HAP problem is also NP-complete. In this chapter, we propose pseudo polynomial time algorithms to optimally solve the HAP prob-lem when the given PDFG is a tree or a simple path, and two other algorithms are proposed to solve the general problem.

Our approach can significantly improve the energy consumption reduction while satisfying the timing constraint. Our contributions are listed as the following:

- When the given PDFG is a tree or a simple path, the results of our algorithms, *Path_Assign* and *Tree_Assign*, cover the results of the optimal *hard HA ILP algorithm* [43] using worst-case scenario of our algorithms.

- For the general problem, that is, when the given input is a *directed acyclic graph* (DAG), our optimal algorithm, *DAG_Opt*, gives the optimal solution and cover the results of the *hard HA ILP algorithm* [43] using worst-case scenario of our algorithms. Our heuristic algorithm, *DAG_Heu*, gives near optimal solutions efficiently.

- Our algorithms are able to give solutions and provide more choices of smaller total costs with guaranteed confidence probabilities satisfying timing constraints. While the *hard HA ILP algorithm* [43] may not find solution with certain timing constraints for hard real-time embedded systems.

- Our algorithms are practical and quick. In practice, when the number of multi-parent nodes and multi-child nodes in the given input graph is small, and the timing constraint is polynomial to the size of PDFG, our algorithms become polynomial. The running times of these algorithms are very small and our experiments always finished in very short time.

We conduct experiments on a set of benchmarks, and compare our algorithms with the *hard HA ILP algorithm* [43]. Experiments show that, when the input PDFG is a tree or a simple path, the results of our algorithms has an average 32.5% improvement with confidence probability 0.9 satisfying timing constraints compared with the results to the hard HA problem. With confidence probability 0.8 satisfying timing constraints, the average improvement is 38.2%; and with confidence probability 0.7 satisfying timing constraints, the improvement is 40.6%. When the input PDFG is a DAG, both our optimal and near optimal algorithms have significant improvement on total cost reduction compared with the cost at hard real-time. On average, our algorithms give a cost reduction of 33.5% with confidence probability 0.9 while satisfying timing constraints, and a cost reduction of 45.3% and 48.9% with confidence probabilities 0.8 and 0.7 while satisfying timing constraints, respectively.

1.2.2 Voltage Assignment with Guaranteed Probability Satisfying Timing Constraint for Real-time Multiproceesor DSP

Energy-saving is a critical issue and performance metric in DSP systems design due to wide use of portable devices, especially those powered by batteries [16,17,28,50,65,101]. The systems become more and more complicate and some tasks may not have fixed execution time. Prior design space exploration methods for hardware/software codesign of DSP systems [42,43,80,95] guarantee no deadline missing by considering worst-case execution time of each task. Dynamic voltage scaling (DVS) is one of the most effective techniques to reduce energy consumption [20,79,82,110].

In Chapter 3, we use probabilistic design space exploration and DVS to avoid over-design systems. We propose two novel optimal algorithms, one for uniprocessor and one for multiprocessor DSP systems, to minimize the expected value of total energy consumption while satisfying timing constraints with guaranteed probabilities for real-time applications. Our work is related to the work in [37, 38]. In [37, 38], Hua et al. proposed an heuristic algorithm for uniprocessor and the *Data Flow Graph* (DFG) is a simple path. We call the offline part of it as *HUA* algorithm for convenience. We also apply the greedy method of *HUA* algorithm to multiprocessor and call the new algorithm as *Heu*.

Our contributions are listed as the following:

- When there is a uniprocessor, the results of our algorithm, *VAP_S*, gives the optimal solution and achieves significant energy saving than *HUA* algorithm.

- For the general problem, that is, when there are multiple processors and the DFG is a DAG, our algorithm, *VAP_M*, gives the optimal solution and achieves significant average energy reduction than *Heu* algorithm.

- Our algorithms not only are optimal, but also provide more choices of smaller expected value of total energy consumption with guaranteed confidence probabilities satisfying timing constraints. In many situations, algorithms *HUA* and *Heu* cannot find a solution, yet ours can find satisfied results.

- Our algorithms are practical and quick. In practice, when the number of multi-parent nodes and multi-child nodes in the given DFG graph is small, and the timing constraint is polynomial to the size of DFG, the running times of these algorithms are very small and our experiments always finished in very short time.

We conduct experiments on a set of benchmarks, and compare our algorithms with *HUA* and *Heu* algorithms Experiments show that our algorithm for uniprocessor, *VAP_S*, has an average 58.0% energy-saving improvement with probability 0.8 satisfying timig constraint compared with the greedy algorithm *HUA*. Our algorithm for multiprocessor and DAG, *VAP_M*, has an average 56.1% energy-saving improvement compared with the results of the heuristic algorithm *Heu*.

1.2.3 Efficient Adaptive Online Energy-Aware Algorithm for Heterogeneous Sensor Networks

Recent advances in heterogeneous wireless communications and electronics have enabled the development of low cost, low power, multifunctional sensor nodes that are small in size and communicate in short distances. These tiny sensor nodes have capability to sense, process data, and communicate [1, 89]. Typically they are densely deployed in large numbers, prone to failures, and their topology changes frequently. They have limited power, computational capacity, bandwidth and memory. As a result of its properties, traditional protocols cannot be applied in this domain [18, 40, 47].

Sensor networks have a wide variety of applications in both military and civil environment. Some of these applications, e.g., natural habitat monitoring, require a large number of tiny sensors and these sensors usually operate on limited battery power. Individual sensors can last only 100-120 hours on a pair of AAA batteries in the active mode. On the other hand, since the number of sensors is huge and they may be deployed in remote, unattended, and hostile environments, it is usually difficult, if not impossible, to recharge or replace their batteries. This problem is compounded by the fact that battery capacity only doubles in 35 years. In a multi-hop ad hoc sensor network, each node plays the dual role of data originator and data router. The malfunctioning of a few nodes can cause significant topological changes and might require rerouting of packets and reorganization of the network [9, 13, 27]. It is for these reasons that researchers are currently focusing on the design of power-aware protocols and algorithms for sensor networks. The main task of a sensor node in a sensor field is to detect events, perform quick local data processing, and then transmit the data. Power consumption can hence be divided into three domains: sensing, communication, and data processing.

Lifetime of distributed micro sensor nodes is a very important issue in the design of sensor networks. The wireless sensor node, being a microelectronic device, can only be equipped with a limited power source (≤ 0.5 Ah, 1.2 V). In some application scenarios, replenishment of power resources might be impossible. Hence, power conservation and power management take on additional importance [46, 102]. Optimizing energy consumption, i.e., minimizing energy consumed by sensing and communication to extend the network lifetime, is an important design objective [84, 105]. To minimize energy consumption and extend network lifetime, a common technique is to put some sensors in the sleep mode and put the others in the active mode for the sensing and communication tasks. When a sensor is in the sleep mode, it is shut down except that a low-power timer is on to wake itself up at a later time [24, 25], therefore it consumes only a tiny fraction of the energy consumed in the active mode [46, 83].

In the data transmission, real-time is a critical requirement for many application for wireless sensor network. There are three modes (active, vulnerable, and sleep) for a sensor network. We call it as *Mode Assignment with Probability* (MAP) problem. For example, in a Bio-sensor, we sample the temperature every minutes. The data collected will go through a fixed topology to the destination. Assume we need the data transmission within 20 seconds. Given this requirement, we need to minimize the total energy consumed in each transmission. Due to the transmission line situation and other overheads, the execution time of each transmission is not a fix number. It may transmit a data in 1 seconds with 0.8 probability and in 3 seconds with 0.2 probability. The mode of a sensor node will affect both the energy and delay of the node.

Sensor networks usually work under dynamic changing and hard-to-predict environments. In this paper, we propose an *adaptive online energy-saving* (AOES) algorithm to reduce the total energy consumption of heterogeneous sensor networks. First, we collect data at certain time interval. During each time interval, the execution time T of each node is estimated by an estimator. We mode T of each node as a random variable and predict the PDF (probability distribution function) of it [37, 38, 92, 111]. Then, we use our optimal *MAP_Opt* algorithm to solve the MAP problem. After finding the best mode assignment for each node, we use dynamic adaptive architecture to adjust the mode of each node online.

For heterogeneous systems [4], each node has different energy consumption rate, which related to area, size, reliability, etc. [26, 32, 42, 52]. Faster one has higher energy consumption while slower one has lower consumption. This paper shows how to assign a proper mode to each node of a *Probability Data Flow Graph* (PDFG) such that the total energy consumption is minimized while the timing constraint is satisfied with a guaranteed confidence probability. With confidence probability P, we can guarantee that the total execution time of the PDFG is less than or equal to the timing constraint with a probability that is greater than or equal to P. In this paper, we compare our optimal sub-algorithm *MAP_Opt* with a heuristic sub-algorithm *MAP_CP*, which is a revised version of previous work. Experiments show significant energy-saving improvement of our algorithm compared with *MAP_CP*.

Our contributions are listed as the following:

- Our algorithm can achieve significant energy-saving for heterogeneous distributed sensor network.

- Our algorithm *MAP_Opt* gives the optimal solution and achieves significant energy saving than *MAP_CP* algorithm.

- Our algorithm *MAP_Opt* not only is optimal, but also provides more choices of smaller total energy consumption with guaranteed confidence probabilities satisfying timing constraints. In many situations, algorithm *MAP_CP* cannot find a solution, while ours can find satisfied results.

- Our algorithm is practical and quick. Extensive experimental results have demonstrated the effectiveness of our approach.

1.2.4 Energy Minimization with Guaranteed Probability Satisfying Timing Constraint via DVS and Loop Scheduling

In Chapter 5, we use probabilistic approach and loop scheduling to avoid over-design systems. We propose a novel optimal algorithm to minimize expected value of total energy consumption while satisfying timing constraints with guaranteed probabilities for real-time applications.

Dynamic voltage scaling (DVS) is one of the most effective techniques to reduce energy consumption [20, 79, 82, 110]. Many researches have been done on DVS for real-time applications in recent years [79, 82, 110]. Zhang et. al. [110] proposed an ILP (Integer Linear Programming) model to solve DVS on multiple processor systems. Shin et. al. [82] proposed a DVS technique for real-time applications based on static timing analysis. However, in the above work, applications are modeled as DAG (Directed Acyclic Graph), and loop optimization is not considered. Saputra et. al. [79] considered loop optimization with DVS. However, in their work, the whole loop is scaled with the same voltage. Our technique can choose the best voltage level assignment for each task node to achieve minimum total energy consumption satisfying the timing constraint.

We design new rotation scheduling algorithms for real-time applications that produce schedules consuming minimal energy. In our algorithms, we use rotation scheduling [15, 17] to get schedules for loop applications. The schedule length will be reduced after rotation. Then, we use DVS to assign voltages to computations individually in order to decrease the voltages of processors as much as possible within the timing constraint. The experimental data show that our algorithms can get better results on energy saving than the previous work. On average, VASP_RS shows a 32.6% reduction in energy consumption compared with the ILP technique in [110] for hard real-time embedded systems. The energy saving results are even better for soft real-time embedded systems.

Our contributions are listed as the following:

- Our algorithm can achieve significant energy-saving than previous work.

- Our algorithm combines both DVS and rotation scheduling to minimize energy consumption.

- Our algorithm *VASP_RS* shows a 32.6% reduction compared with the ILP technique in [110] for hard real-time and even better for soft real-time.

1.2.5 Loop Scheduling to Minimize Cost with Data Mining and Prefetching for Heterogeneous DSP

In high level synthesis, cost (such as energy, reliability, etc.) minimization has become a primary concern in today's real-time embedded systems. In DSP systems, some tasks may not have fixed execution time. Such tasks usually contain conditional instructions and/or operations that could have different execution times for different inputs. It is possible to obtain the execution time distribution for each task by sampling or profiling [91]. In this paper, we use a data mining engine to do the prediction. First, the data mining engine collects data into the log. Then do clustering and use unsupervising method to find the distribution pattern of all random variables, i.e., the execution times. Finally, the engine builds the distribution function for each execution time that has uncertainty.

Our approach includes three steps: First, we use data mining to predict the execution time pattern and time-input association from history table. Second, we use rotation scheduling to obtain the best assignment for total cost minimization. This includes finding the best FU type assignment in each iteration and rotation scheduling for Q iterations. Finally, we use prefetching to prepare data in advance at run time. The experimental data shows that our algorithms can get better results on cost saving than the previous work.

Data mining engine works at compile time. Based on the obtained distribution function of each execution time, we will use loop scheduling to find the best assignment for cost minimization. Then at run time, needed data will be prefetched in advance. Based on the computed best assignment, we can prefetch data in certain time ahead with guaranteed probability. For example, if node A select type $F1$, then we prefetch data for node B in 2 time unit in advance, and it will guarantee with 100% that node B can be executed on time with needed data.

We have several contributions in this novel approach:

- By taking advantage of the uncertainties in execution time of tasks and using data mining, we give out FU assignments and scheduling to minimize the expected total cost while satisfying timing constraint with guaranteed probabilities.

- We repeatedly regroup a loop based on rotation scheduling and decrease the energy by voltage selection as much as possible within a timing constraint.

- We prefetch the data needed in advance at run time. Our approach can handle loops efficiently.

1.2.6 Type Assignment and Scheduling with Variable Partition to Minimize Energy for Heterogeneous Multi-Bank Memory

Memory access time latency and energy consumption are two of the most important design considerations in memory architecture. A number of papers have investigated how to exploit multi-bank memory from either performance or energy aspect. But the combined effect of both performance and energy requirement is seldom tackled. In high-performance *digital signal processing* (DSP) applications, strict real-time processing is critical [112] since the growing speed gap between *central processing unit* (CPU) and memory becomes a bottleneck for designing such real-time systems. To close this speed gap, embedded systems need to utilize multi-bank on-chip memories [53, 98, 99]. The high energy consumption of memories make them target of many energy-conscious optimization techniques [12]. This is especially true for mobile applications which are typically memory-intensive, such as signal and video processing. This chapter focuses on reducing the total energy while satisfying performance constraints on multi-bank memory architectures.

Combining the consideration of energy and performance, in Chapter 7, we propose a novel memory model to overcome the weakness of previous techniques. An algorithm, TAMRS (*Type Assignment and Minimum Resource Scheduling*), is proposed. The algorithm use type assignment with the consideration of variable partition to find the best configuration for both memory and ALU.

The experimental results show that TAMRS achieves a significant reduction on average in total energy consumption. For example, with 3 memory types and 3 ALU types, compared with the VPIS algorithm in Wang et al.'s paper [97], TAMRS shows an average 15.6% reduction in total energy consumption.

In summary, in Chapter 7, we have several major contributions:

- We study the combined effects of energy-saving and performance of memory and ALU in a systematic approach.

- We exploit the energy saving of memory with operation scheduling and memory type assignment.

- Data locality has been improved by using variable partition.

1.3 Contributions

In this dissertation, our research focuses on understanding fundamental properties and developing models, methodologies, and algorithms for power-aware high-performance embedded systems. We have performed our research from various aspects including high-level architecture synthesis, low power design, and power & performance optimization for memory. A lot of promising results in these fields have been yielded, and these results tremendously improve the state-of-the-art techniques. Our contributions are summarized as follows:

1. We propose a theoretical foundation for an important problem in high-level architecture synthesis for soft real-time DSP using heterogeneous functional units (FUs), *heterogeneous assignment with probability* (HAP) problem. We model the execution time of each node as a random variable. We use several efficient algorithms to assign a proper FU type to each operation of a DSP application in such a way that all requirements can be met and the total cost can be minimized while satisfying timing constraints with guaranteed probabilities.

2. The solutions to the HAP problem are useful for both hard real-time and soft real-time systems. Optimal algorithms are proposed to find the optimal solutions for the HAP problem when the input is a tree or a simple path. Two other algorithms, one is optimal and the other is near-optimal heuristic, are proposed to solve the general problem. The experiments demonstrated the effectiveness of our algorithms.

3. We propose an novel algorithm which combine *Dynamic Voltage Scaling* (DVS) and soft real-time to reduce energy consumption of uniprocessor and multiprocessor by solving *Voltage Assignment with Probability* (VAP) problem. VAP problem involves finding a voltage level to be used for each node of an date flow graph (DFG) in uniprocessor and multiprocessor DSP systems.

4. We propose two optimal algorithms, one for uniprocessor and one for multiprocessor DSP systems, to minimize the expected total energy consumption while satisfying the timing constraint with a guaranteed confidence probability. The experimental results show that our approach achieves significant energy saving than previous work. For example, our algorithm for multiprocessor achieves an average improvement of 56.1% on total energy-saving with 0.80 probability satisfying timing constraint.

5. We study the energy saving issue in heterogeneous sensor networks. As we know, energy and timing are critical issues for wireless sensor networks since most sensors are equipped with non-rechargeable batteries that have limited lifetime. However, sensor nodes usually work under dynamic changing and hard-to-predict environments. We proposed a novel *adaptive online energy-saving* (AOES) algorithm to save total energy consumption for heterogeneous sensor networks.

6. Due to the uncertainties in execution time of some tasks and multiple working mode of each node, In AOES algorithm, we model each varied execution time as a probabilistic random variable, and saved energy by selecting the best mode assignment for each node, which is called MAP (Mode Assignment with Probability) problem. We propose an optimal sub-algorithm *MAP_Opt* to minimize the total energy consumption while satisfying the timing constraint with a guaranteed confidence probability.

7. We use loop scheduling to further extend our work of VAP problem. Low energy consumption is an important problem in real-time embedded systems and loop is the most energy consuming part in most cases. Due to the uncertainties in execution time of some tasks, this chapter models each varied execution time as a probabilistic random variable. We use rotation scheduling and DVS (Dynamic Voltage Scaling) to minimize the expected total energy consumption while satisfying the timing constraint with a guaranteed confidence probability.

8. By using loop scheduling, our approach can handle loops efficiently. In addition, it is suitable to both soft and hard real-time systems. And even for hard real-time, we have good results. The experimental results show that our approach achieves significant energy saving than list scheduling and ILP (Integer Linear Programming) voltage assignment.

9. We combine data mining and prefetching to reduce energy consumptions. The basic steps are as follows: First, we use data mining to predict the distribution of execution time and find the association rules between execution time and different inputs from history table. Then we use rotation scheduling to obtain the best assignment for total cost minimization. Finally, we use prefetching to prepare data in advance at run time. Experiments demonstrate the effectiveness of our algorithm. Our approach can handle loops efficiently. In addition, it is suitable to both soft and hard real-time systems.

10. We address a critical problem in multi-bank on-chip memory. In many high-performance DSP processors, multi-bank on-chip memory was employed to improve performance and energy consumption. This architectural feature supports higher memory bandwidth by allowing multiple data memory accesses to be executed in parallel. However, making effective use of multi-bank memory remains difficult, considering the combined effect of performance and energy requirement.

11. We study the scheduling and assignment problem that minimizes the total energy while satisfying performance requirements. An algorithm, TAMRS (*Type Assignment and Minimum Resource Scheduling*), is proposed. The algorithm attempts to maximum energy-saving while satisfying timing constraints. The experimental results show that the average improvement on energy-saving is significant by using TAMRS.

1.4   Outline

The rest of this dissertation is organized as follows: In Chapter 2, we address high-level synthesis for real-time digital signal processing (DSP) architectures using heterogeneous functional units (FUs). In high-level synthesis for real-time embedded systems using heterogeneous functional units (FUs), it is critical to select the best FU type for each task. However, some tasks may not have fixed execution times. Chapter 2 models each varied execution time as a probabilistic random variable and solves *heterogeneous assignment with probability* (HAP) problem. The solution of the HAP problem assigns a proper FU type to each task such that the total cost is minimized while the timing constraint is satisfied with a guaranteed confidence probability. The solutions to the HAP problem are useful for both hard real-time and soft real-time systems. Optimal algorithms are proposed to find the optimal solutions for the HAP problem when the input is a tree or a simple path. Two other algorithms, one is optimal and the other is near-optimal heuristic, are proposed to solve the general problem. The experiments show that our algorithms can effectively reduce the total cost while satisfying timing constraints with guaranteed confidence probabilities. For example, our algorithms achieve an average reduction of 33.0% on total cost with 0.90 confidence probability satisfying timing constraints compared with the previous work using worst-case scenario.

In Chapter 3, we propose a novel algorithm which combine *Dynamic Voltage Scaling* (DVS) and soft real-time to reduce energy consumption of uniprocessor and multiprocessor. DVS is one of the techniques used to obtain energy-saving in real-time DSP systems. In many DSP systems, some tasks contain conditional instructions that have different execution times for different inputs. Due to the uncertainties in execution time of these tasks, this chapter models each varied execution time as a probabilistic random variable and solves the *Voltage Assignment with Probability* (VAP) Problem. VAP problem involves finding a voltage level to be used for each node of an date flow graph (DFG) in uniprocessor and multiprocessor DSP systems. This chapter proposes two optimal algorithms, one for uniprocessor and one for multiprocessor DSP systems, to minimize the expected total energy consumption while satisfying the timing constraint with a guaranteed confidence probability. The experimental results show that our approach achieves significant energy saving than previous work. For example, our algorithm for multiprocessor achieves an average improvement of 56.1% on total energy-saving with 0.80 probability satisfying timing constraint.

In Chapter 4, we study the energy saving issue in heterogeneous sensor networks. As we know, energy and timing are critical issues for wireless sensor networks since most sensors are equipped with non-rechargeable batteries that have limited lifetime. However, sensor nodes usually work under dynamic changing and hard-to-predict environments. This chapter uses a novel *adaptive online energy-saving* (AOES) algorithm to save total energy consumption for heterogeneous sensor networks. Due to the uncertainties in execution time of some tasks and multiple working mode of each node, this paper models each varied execution time as a probabilistic random variable to save energy by selecting the best mode assignment for each node, which is called MAP (Mode Assignment with Probability) problem in this paper. We propose an optimal sub-algorithm *MAP_Opt* to minimize the total energy consumption while satisfying the timing constraint with a guaranteed confidence probability.

In Chapter 5, we use loop scheduling to further extend our work in Chapter 3. Low energy consumption is an important problem in real-time embedded systems and loop is the most energy consuming part in most cases. Due to the uncertainties in execution time of some tasks, this chapter models each varied execution time as a probabilistic random variable. We use rotation scheduling and DVS (Dynamic Voltage Scaling) to minimize the expected total energy consumption while satisfying the timing constraint with a guaranteed confidence probability. Our approach can handle loops efficiently. In addition, it is suitable to both soft and hard real-time systems. And even for hard real-time, we have good results. The experimental results show that our approach achieves significant energy saving than list scheduling and ILP (Integer Linear Programming) voltage assignment.

We combine loop scheduling and data mining and prefetching to solve the heterogeneous FU assignment problem, which is shown in Chapter 6. We propose an algorithm, *LSHAP*, to give the efficient solutions. We first use data mining to obtain the PDF of execution times, which are modeled as random variable in this paper. And find the association rules between execution times and different inputs. Then, by taking advantage of the uncertainties in execution time of tasks, we give out FU assignments and scheduling to minimize the expected total cost while satisfying timing constraint with guaranteed probabilities. We repeatedly regroup a loop based on rotation scheduling and decrease the energy by voltage selection as much as possible within a timing constraint. Finally, we prefetch the data needed in advance at run time. Our approach can handle loops efficiently. Experimental results show the significant cost-saving of our approach.

In Chapter 7, we address a critical problem in multi-bank on-chip memory. In many high-performance DSP processors, multi-bank on-chip memory was employed to improve performance and energy consumption. This architectural feature supports higher memory bandwidth by allowing multiple data memory accesses to be executed in parallel. However, making effective use of multi-bank memory remains difficult, considering the combined effect of performance and energy requirement. This chapter studies the scheduling and assignment problem that minimizes the total energy while satisfying performance requirements. An algorithm, TAMRS (*Type Assignment and Minimum Resource Scheduling*), is proposed. The algorithm attempts to maximum energy-saving while satisfying timing constraints. The experimental results show that the average improvement on energy-saving is significant by using TAMRS.

Finally, in Chapter 8, we provide concluding remarks for our research work presented in this dissertation. Most of them have already been reported to research community through publications [66–76]. A brief discussion on future work is also provided in this chapter.

CHAPTER 2

HETEROGENEOUS ASSIGNMENT TO MINIMIZE COST WHILE SATISFYING HARD/SOFT TIMING CONSTRAINTS

In high-level synthesis for real-time embedded systems using heterogeneous functional units (FUs), it is critical to select the best FU type for each task. However, some tasks may not have fixed execution times. This chapter models each varied execution time as a probabilistic random variable and solves *heterogeneous assignment with probability* (HAP) problem. The solution of the HAP problem assigns a proper FU type to each task such that the total cost is minimized while the timing constraint is satisfied with a guaranteed confidence probability. The solutions to the HAP problem are useful for both hard real-time and soft real-time systems. Optimal algorithms are proposed to find the optimal solutions for the HAP problem when the input is a tree or a simple path. Two other algorithms, one is optimal and the other is near-optimal heuristic, are proposed to solve the general problem. The experiments show that our algorithms can effectively reduce the total cost while satisfying timing constraints with guaranteed confidence probabilities. For example, our algorithms achieve an average reduction of 33.0% on total cost with confidence probability 0.90 satisfying timing constraints compared with the previous work using worst-case scenario.

2.1 Introduction

It is a critical issue to do high level synthesis for special purpose architectures of real-time embedded systems to satisfy the time and cost requirements [16, 17, 28, 42, 43, 50,

65, 95, 101]. The cost of embedded systems may relate to power, reliability, etc. [26, 32, 81, 86]. The systems become more and more complicate in two aspects. First, in many systems, such as heterogeneous parallel DSP systems [4, 33], same type of operations can be processed by heterogeneous FUs with different costs. Therefore, an important problem arises: how to assign a proper FU type to each operation of a DSP application such that the requirements can be met and the total cost can be minimized while satisfying timing constraints with a guaranteed confidence probability [80].

Second, some tasks may not have fixed execution time. Such tasks usually contain conditional instructions and/or operations that could have different execution times for different inputs [36–38, 92, 111]. Although many static assignment techniques can thoroughly check for the best assignment for dependent tasks, existing methods are not able to deal with such uncertainty. Therefore, either worst-case or average-case computation times for these tasks are usually assumed. Such assumptions, however, may not be applicable for real-time systems and may result in an uneffective task assignment. Using probabilistic approach, we can obtain solutions that can not only be used for hard real-time systems, but also provide more choices of smaller total costs while satisfying timing constraints with guaranteed confidence probabilities.

This chapter presents high-level synthesis algorithms which operate in probabilistic environments to solve the *heterogeneous assignment with probability* (HAP) problem. In the HAP problem, we model the execution time of a task as a random variable [108]. For heterogeneous systems, each FU type has different cost, representing hardware cost, size, reliability, etc. Faster one has higher cost while slower one has lower cost. This chapter shows how to assign a proper FU type to each node of a *Probabilistic Data Flow Graph* (PDFG) such that the total cost is minimized while satisfying the timing constraint with a guaranteed confidence probability. In other words, we can guarantee that the total execution time of the PDFG is less than or equal to the timing constraint with a probability greater than or equal to P.

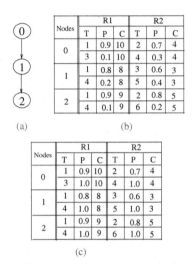

(a)

Nodes	R1			R2		
	T	P	C	T	P	C
0	1	0.9	10	2	0.7	4
	3	0.1	10	4	0.3	4
1	1	0.8	8	3	0.6	3
	4	0.2	8	5	0.4	3
2	1	0.9	9	2	0.8	5
	4	0.1	9	6	0.2	5

(b)

Nodes	R1			R2		
	T	P	C	T	P	C
0	1	0.9	10	2	0.7	4
	3	1.0	10	4	1.0	4
1	1	0.8	8	3	0.6	3
	4	1.0	8	5	1.0	3
2	1	0.9	9	2	0.8	5
	4	1.0	9	6	1.0	5

(c)

Figure 2.1. (a) A given simple path (b) The times, probabilities, and costs of its node for different FU types. (c) The time cumulative distribution functions (CDFs) and costs of its node for different FU types.

We show an example to illustrate the HAP problem. Assume that the FU type library provides two types of FUs, R_1 and R_2, for us to select from. An exemplary PDFG is shown in Figure 2.1(a), which is a simple path with 3 nodes. The execution times (T), probabilities (P), and costs (C) of each node for different FU types are shown in Figure 2.1(b). Each node can select one of the two different FU types. The execution time (T) of each FU type is modeled as a random variable. The probabilities may come from a statistical profiling. For example, node 0 can choose one of the two types: R_1 or R_2. When choosing R_1, node 0 will be finished in 1 time unit with probability 0.9 and will be finished in 3 time units with probability 0.1. In other words, node 0 can guarantee to be finished in 3 time units with 100% probability. Hence, we care about the time *cumulative distribution function* (CDF) $F(t)$, which gives accumulated probability for $T \leq t$. For example, the CDF of node 0 at time units 1 is 0.9. And the CDF of node 0 at time units 3 is 1.0. Figure 2.1(c) shows the time CDFs and costs of each node for different FU types.

A solution to the HAP problem with timing constraint 10 can be found as follows: We assign FU types 2, 2, and 1 for nodes 0, 1, and 2, respectively. Let T_0, T_1, and T_2 be the random variables representing the execution times of nodes 0, 1, and 2. From Figure 2.1(c), we get: $Pr(T_0 \leq 4) = 1.0$, $Pr(T_1 \leq 5) = 1.0$, and $Pr(T_2 \leq 1) = 0.9$. Hence, we obtain minimum total cost 16 with 0.9 probability satisfying the timing constraint 10. The total cost is computed by adding the costs of all nodes together and the probability corresponding to the total cost is computed by multiplying the probabilities of all nodes based on the basic properties of probability and cost of a PDFG.

In Figure 2.1(b), if we use the worst-case execution time as a fixed execution time for each node, then the assignment problem becomes the *hard heterogeneous assignment* (hard HA) problem, which is related to the hard real-time. The hard HA problem is the worst-case scenario of the *heterogeneous assignment with probability* (HAP) problem. For example, in the hard HA problem, when choosing type R_1, node 0 has only one execution time 3. When choosing type R_2, node 0 has one execution time 4. With certain timing constraints, there might not be solution for the hard HA problem. However, for soft real-time applications, it is desirable to find an assignment that guarantees the total execution time to be less than or equal to the timing constraint with certain confidence probability.

For example, in Figure 2.1, under timing constraint 10, we cannot find a solution to the hard HA problem. But we can obtain minimum system cost 16 with probability 0.9 satisfying the timing constraint 10. Also, the cost obtained from worst-case scenario is always larger than or equal to the cost from the probabilistic scenario. For example, under timing constraint 11, the minimum cost is 27 for the hard HA problem. While in the HAP problem, with confidence probability 0.9 satisfying the timing constraint we get the minimum cost of 16, which gives 40.7% improvement.

It is known that the hard HA problem is NP-complete [80]. Since the HAP problem is NP harder than the hard HA problem, the HAP problem is also NP-complete. In this chapter, we propose pseudo polynomial time algorithms to optimally solve the HAP problem when the given PDFG is a tree or a simple path, and two other algorithms are proposed to solve the general problem.

There have been a lot of research efforts on allocating applications in heterogeneous distributed systems [4–8, 10, 33, 77]. Incorporating reliability cost into heterogeneous distributed systems, the reliability driven assignment problem has been studied in [26, 81, 86]. In these work, allocation are performed based on a fixed architecture. However, when performing assignment in architecture synthesis, no fixed architectures are available. Most previous work on the synthesis of special purpose architectures for real-time DSP applications focuses on the architectures that only use homogeneous FUs, that is, same type of operations will be processed by same type of FUs [14, 17, 29, 39, 42, 52, 56, 57, 62, 65].

Our work is related to the work in [43, 80, 92]. *Probabilistic retiming* (PR) had been proposed in [64, 92]. For a system without resource constraints, PR can be applied to optimize the input graph, i.e., reduce the length of the longest path of the graph such that the probability of the longest path computation time being less than or equal to the given timing constraint, L, is greater than or equal to a given confidence probability P. Since the execution times of the nodes can be either fixed or varied, a probability model is employed to represent the execution time of the tasks. But PR does not model the hard HA problem which focuses on how to obtain the best assignment from different FU types. In [80], the authors proposed two optimal algorithms for the hard HA problem when the given input is a tree or simple path, and three heuristic algorithms for the general hard HA problem. But they do not consider varied execution time situation. Also, for the general problem, their solutions are not optimal. Although the ILP model from [43] can obtain an optimal solution for the heterogeneous assignment problem, it is a NP-hard problem to solve the ILP model. Therefore, the ILP model may take a very long time to get results even when a given *Data Flow Graph* (DFG) is not very big. In this chapter, the algorithm solving the hard HA problem in [43] is called *hard HA ILP algorithm* [43] in general.

Our contributions are listed as the following:

- When the given PDFG is a tree or a simple path, the results of our algorithms, *Path_Assign* and *Tree_Assign*, cover the results of the optimal *hard HA ILP algorithm* [43] using worst-case scenario of our algorithms. And our algorithms produce results for soft real-time systems.

- For the general problem, that is, when the given input is a *directed acyclic graph* (DAG), our optimal algorithm, *DAG_Opt*, gives the optimal solution and cover the results of the *hard HA ILP algorithm* [43] using worst-case scenario of our algorithms. Our heuristic algorithm, *DAG_Heu*, gives near optimal solutions efficiently.

- Our algorithms are able to give solutions and provide more choices of smaller total costs with guaranteed confidence probabilities satisfying timing constraints. While the *hard HA ILP algorithm* [43] may not find solution with certain timing constraints.

- Our algorithms are practical and quick. In practice, when the number of multi-parent nodes and multi-child nodes in the given input graph is small, and the timing constraint is polynomial to the size of PDFG, our algorithms become polynomial. The running times of these algorithms are very small and our experiments always finished in very short time.

We conduct experiments on a set of benchmarks, and compare our algorithms with the *hard HA ILP algorithm* [43]. Experiments show that, when the input PDFG is a tree or a simple path, the results of our algorithms has an average 32.5% improvement with 0.9 confidence probability satisfying timing constraints compared with the results to the hard HA problem. With 0.8 confidence probability satisfying timing constraints, the average improvement is 38.2%; and with 0.7 confidence probability satisfying timing constraints, the improvement is 40.6%. When the input PDFG is a DAG, both our optimal and near optimal algorithms have significant improvement on total cost reduction compared with the cost at hard real-time. On average, our algorithms give a cost reduction of 33.5% with 0.9 confidence probability satisfying timing constraints, and a cost reduction of 45.3% and 48.9% with 0.8 and 0.7 confidence probability satisfying timing constraints, respectively.

The remainder of this chapter is organized as follows: In the next section, we give the basic definitions and models used in the rest of the chapter. Examples of the HAP problem when the input is a simple path or a tree are given in Section 2.3, The algorithms for HAP problem are presented in Section 2.4. Experimental results and concluding remarks are provided in Section 2.5 and Section 2.6 respectively.

2.2 System Model

We use *Probabilistic Data Flow Graph (PDFG)* to model a application of embedded systems. A **PDFG G** $= \langle V, E, T, R \rangle$ is a *directed acyclic graph* (DAG), where $V = \langle v_1, v_2, \cdots, v_N \rangle$ is the set of nodes, $E \subseteq V \times V$ is the edge set that defines the precedence relations among nodes in V. In practice, many architectures consist of different types of FUs. Assume there are maximum M different FU types in a FU set R=$\{R_1, R_2, \cdots, R_M\}$. For each FU type, there are maximum K execution time variations T, although each node may have different number of FU types and execution time variations.

An assignment for a PDFG G is to assign a FU type to each node. Define an **assignment A** to be a function from domain V to range R, where V is the node set and R is FU type set. For a node $v \in V$, $A(v)$ gives the selected type of node v. For example, in Figure 2.1(a), Assigning FU types 2, 2, and 1 for nodes 0, 1, and 2, respectively, we obtain minimum total cost 16 with 0.9 probability satisfying the timing constraint 10. That is, $A(0) = 2$, $A(1) = 2$, and $A(2) = 1$.

In a PDFG G, each varied execution time T is modeled as a probabilistic random variable. $\mathbf{T_{R_j}}(\mathbf{v})$ $(1 \leq j \leq M)$ represents the execution times of each node $v \in V$ for FU type j, and $\mathbf{P_{R_j}}(\mathbf{v})$ $(1 \leq j \leq M)$ represents the corresponding probability function. And $\mathbf{C_{R_j}}(\mathbf{v})$ $(1 \leq j \leq M)$ is used to represent the cost of each node $v \in V$ for FU type j, which is a fixed value. For instance, in Figure 2.1(a), $T_1(0) = 1, 3$; $T_2(0) = 2, 4$. Correspondingly, $P_1(0) = 0.9, 0.1$; $P_2(0) = 0.7, 0.3$. And $C_1(0) = 10$; $C_2(0) = 4$.

Given an assignment A of a PDFG G, we define the **system total cost under assignment A**, denoted as $\mathbf{C_A(G)}$, to be the summation of costs of all nodes, that is, $C_A(G) = \sum_{v \in V} C_{A(v)}(v)$. In this chapter we call $C_A(G)$ as **total cost** in brief. For example, in Figure 2.1(a), under assignment 2, 2, and 1 for nodes 0, 1, and 2, respectively, the costs of nodes 0, 1, and 2 are: $C_2(0) = 4$, $C_2(1) = 3$, and $C_1(3) = 9$. Hence, the total cost of the graph G is: $C_A(G) = C_2(0) + C_2(1) + C_1(3)$, that is, $C_A(G) = 16$.

For the input PDFG G, given an assignment A, assume that $\mathbf{T_A(G)}$ stands for the **execution time of graph G under assignment A**. $T_A(G)$ can be gotten from the longest path p in G. The new variable $T_A(G) = \max_{\forall p} T_{A(v)}(p)$, where $T_{A(v)}(p) = \sum_{v \in p} T_{A(v)}(v)$, is also a random variable. In Figure 2.1(a), there is only one path. Under assignment 2, 2, and 1 for nodes 0, 1, and 2, $T_A(G) = T_{A(v)}(p) = T_2(0) + T_2(1) + T_1(3)$. Since $T_2(0)$, $T_2(1)$, and $T_1(3)$ all are random variables, and the summation of random variables is also a random variable, $T_A(G)$ is also a random variable.

The *minimum total cost C with confidence probability P under timing constraint L* is defined as $C = \min_A C_A(G)$, where probability of $(T_A(G) \leq L) \geq P$. Probability of ($T_A(G) \leq L$) is computed by multiplying the probabilities of all nodes together while satisfying $T_A(G) \leq L$. That is, $P_A(G) = \prod_{v \in V} P_{A(v)}(v)$.

In Figure 2.1(a), under assignment 2, 2, and 1 for nodes 0, 1, and 2, $P_2(0) = Pr(T_2(0) \leq 4) = 1.0$, $P_2(1) = Pr(T_2(1) \leq 5) = 1.0$, and $P_1(3) = Pr(T_1(3) \leq 1) = 0.9$. Hence, $P_A(G) = \prod_{v \in V} P_{A(v)}(v) = 0.9$. With confidence probability P, we can guarantee that the total execution time of the graph G is less than or equal to the timing constraint L with a probability greater than or equal to P. For each timing constraint L, our algorithm will output a serial of (Probability, Cost) pairs (P, C).

$T_{R_j}(v)$ is either a discrete random variable or a continuous random variable. We define **F** to be the *cumulative distribution function* of the random variable $T_{R_j}(v)$ (abbreviated as **CDF**), where $F(t) = P(T_{R_j}(v) < t)$. When $T_{R_j}(v)$ is a discrete random variable, the CDF $F(t)$ is the sum of all the probabilities associating with the execution times that are less than or equal to t. Figure 2.1(c) gives the time CDFs of each node for different FU types. If $T_{R_j}(v)$ is a continuous random variable, then it has a *probability density function (PDF)* . If assume the PDF is f, then $F(t) = \int_0^t f(s)ds$. Function F is non-decreasing, and $F(-\infty) = 0$, $F(\infty) = 1$.

We define the *heterogeneous assignment with probability (HAP)* problem as follows: Given M different Functional Unit types: $R_1, R_2, \cdots, R_M$, a PDFG $G = \langle V, E \rangle$ where $V = \langle v_1, v_2, \cdots, v_N \rangle$, $T_{R_j}(v)$, $P_{R_j}(v)$, $C_{R_j}(v)$ for each node $v \in V$ executed on each FU type j, and a timing constraint L, find an assignment for G that gives the *minimum total cost C with confidence probability P under timing constraint L*. In Figure 2.1(a), a solution to the HAP problem with $T = 11$ can be found as follows. Assigning FU types 2, 2, and 2 for nodes 0, 1, and 2, respectively, we obtain minimum total cost 12 with 0.8 probability.

2.3 Examples

Here we give two different types of examples to illustrate the *heterogeneous assignment probability* (HAP) problem. One is a simple path, and the other is a tree for the given input PDFG.

2.3.1 Simple Path

For the example in Figure 2.1(a), each node has two different FU types to choose from, and is executed on them with probabilistic times. In many applications, a real-time system does not always has hard deadline time. The execution time can be smaller than the hard deadline time with certain probabilities. So, the hard deadline time is the worst-case of the varied smaller time cases. If we consider these time variations, we can achieve a better minimum cost with satisfying confidence probabilities.

Table 2.1. Minimum total costs with computed confidence probabilities under various timing constraints for a simple path

T	(P , C)	(P , C)	(P , C)	(P , C)	(P , C)	(P , C)
3	0.65, 27					
4	0.50, 21	0.58, 23	0.65, 27			
5	0.45, 17	0.50, 21	0.58, 23	0.72, 27		
6	0.38, 16	0.45, 17	0.72, 21	0.81, 27		
7	0.34, 12	0.38, 16	0.64, 17	0.72, 21	0.81, 22	
8	0.34, 12	0.63, 16	0.64, 17	0.72, 18	0.81, 22	0.90, 27
9	0.56, 12	0.63, 16	0.64, 17	0.72, 18	0.90, 21	
10	0.56, 12	0.90, 16				
11	0.80, 12	0.90, 16	**1.00, 27**			
12	0.80, 12	0.90, 16	**1.00, 21**			
13	0.80, 12	**1.00, 16**				
14	0.80, 12	**1.00, 16**				
15	**1.00, 12**					

Different FU types have different costs, which can be any cost such as hardware cost, energy consumption or reliability cost. A node may run slower but with less energy consumption or reliability cost when executed on one type of FUs than on another [80]. But for the same FU type, a node may have varied execution times while there is not too much difference in cost. In this paper, we assume the cost of a FU type is fixed for a node executed on it while the execution time is a random variable. When the cost is related to energy consumption, it is clear that the total energy consumption is the summation of energy cost of each node. Also, when the cost is related to reliability, the total reliability cost is the summation of reliability cost of all nodes. We compute the reliability cost using the same model as in [86]. From the conclusion of papers [86] and [80], we know that in order to maximize the reliability of a system, we need to find an assignment such that the timing constraint is satisfied and the summation of reliability costs of all nodes is minimized.

For this simple path with three nodes, we can obtain the minimum total cost with computed confidence probabilities under various timing constraints. The results generated by our algorithms are shown in Table 2.1. The entries with probability equal to 1 (see the entries in boldface) actually give the results to the hard HA problem which show the worst-case scenario of the HAP problem. For each row of the table, the C in each (P, C) pair gives the minimum total cost with confidence probability P under timing constraint T. For example, when $T = 3$, with pair (0.65, 27), we can achieve minimum total cost 27 with confidence probability 0.65 under timing constraint 3.

Comparing with optimal results of the hard HA problem for a simple path using worst-case scenario, Table 2.1 provides more information, more selections and decisions, no matter whether the system is hard or soft real-time. In Table 2.1, we have the output of our algorithm from timing constraint 3 to 15, while the optimal results of the hard HA problem for a simple path only has 5 entries (in boldface) from timing constraint 11 to 15.

For a soft real-time system, some nodes of PDFG have smaller probabilistic execution times compared with the hard deadline time. We can achieve much smaller cost than the cost of worst-case with guaranteed confidence probability. For example, under timing

Table 2.2. With timing constraint 11, the assignments of types for each node with different (Probability, Cost) pairs.

	Node id	Type id	T	Prob.	Cost
Assign	0	2	4	1.00	4
Assign	1	2	5	1.00	3
Assign	2	2	2	0.80	5
Total			11	0.80	12
Assign	0	2	4	1.00	4
Assign	1	2	5	1.00	3
Assign	2	1	1	0.90	9
Total			11	0.90	16
Assign	0	1	3	1.00	10
Assign	1	1	4	1.00	8
Assign	2	1	4	1.00	9
Total			11	1.00	27

constraint 11, we can select the pair (0.90, 16) which guarantees to achieve minimum cost 16 with 0.90 confidence satisfying the timing constraint 11. It achieves 40.7% reduction in cost compared with the cost 27, the result obtained by the algorithms using worst-case scenario of the HAP problem. In many situations, this assignment is good enough for users. We also can select the pair (0.80, 12) which has provable confidence probability 0.8 satisfying the timing constraint 11, while the cost 12 is only 55.5% of the cost of worst-case, 27. The assignments for each pair of (0.80, 12), (0.90, 16), and (1.00, 27) under the timing constraint 11 are shown in Table 2.2.

Table 2.3. Given an assignment for simple path, the (Probability, Cost) pairs under different timing constraints.

T	6, 7	8 , 9	10 - 12	13
(P , C)	(0.38, 16)	(0.63, 16)	(0.90, 16)	(1.00, 16)

Given an assignment, we can get a minimum total cost under every timing constraint. But the probability to achieve this minimum total cost may not be same. For example, if the assignments for nodes 0, 1, and 2 are types 2, 2, and 1, respectively, then the total cost is 12. But the probabilities vary from 0.38 to 1.00. The probabilities under different timing constraints are shown in Table 2.3.

2.3.2 Tree

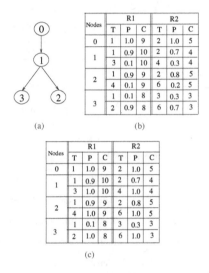

(b)

Nodes	R1			R2		
	T	P	C	T	P	C
0	1	1.0	9	2	1.0	5
1	1	0.9	10	2	0.7	4
	3	0.1	10	4	0.3	4
2	1	0.9	9	2	0.8	5
	4	0.1	9	6	0.2	5
3	1	0.1	8	3	0.3	3
	2	0.9	8	6	0.7	3

(c)

Nodes	R1			R2		
	T	P	C	T	P	C
0	1	1.0	9	2	1.0	5
1	1	0.9	10	2	0.7	4
	3	1.0	10	4	1.0	4
2	1	0.9	9	2	0.8	5
	4	1.0	9	6	1.0	5
3	1	0.1	8	3	0.3	3
	2	1.0	8	6	1.0	3

Figure 2.2. (a) A given tree. (b) The times, probabilities, and costs of its nodes for different FU types. (c) The time cumulative distribution functions (CDFs) and costs of its node for different FU types.

Here we give an example of the HAP problem when the input is a tree. The PDFG graph is shown in Figure 2.2(a), which is a tree with four nodes. The times, costs, and probabilities are shown in Figure 2.2(b). For example, node 1 can choose one of the two types: R_1 or R_2. When choosing R_1, node 1 will be finished in 1 time unit with probability 0.9 and will be finish in time units 3 with probability 0.1. The cost of type R_1 is 10. When node 1 choose type R_2, it will be finished within 2 time units with probability 0.7 and will be finished in 4 time units with probability 0.3. The cost of type R_2 is 4. Node 0

45

has two types of FUs to execute on. But for each type, node 0 has fixed execution time. Figure 2.2(c) shows the time cumulative distribution functions (CDFs) and costs of each node for different FU types.

Table 2.4. Minimum total costs with computed confidence probabilities under various timing constraints for a tree.

T	(P , C)	(P , C)	(P , C)	(P , C)	(P , C)	(P , C)
3	0.08, 36					
4	0.06, 30	0.72, 32	0.81, 36			
5	0.56, 26	0.72, 28	0.81, 32			
6	0.17, 21	0.56, 22	0.63, 26	0.72, 28	0.81, 32	0.90, 36
7	0.17, 17	0.19, 21	0.56, 22	0.80, 26	0.90, 30	
8	0.17, 17	0.24, 21	0.80, 22	0.90, 26	**1.00, 36**	
9	0.24, 17	0.70, 21	0.80, 22	0.90, 23	**1.00, 30**	
10	0.70, 17	0.80, 22	0.90, 23	**1.00, 26**		
11	0.70, 17	**1.00, 21**				
12	**1.00, 17**					

The minimum total costs with computed confidence probabilities under various timing constraints for a tree are shown in Table 2.4. The entries with probability equal to 1 (see the entries in boldface) actually give the results to the hard HA problem which shows the worst-case scenario of the HAP problem.

From above two examples, we can see that the probabilistic approach to heterogeneous assignment problem has great advantages: It provides the possibility to reduce the total costs of systems under different timing constraints with guaranteed confidence probabilities. It is suitable to both hard and soft real-time systems. We will give related algorithms and experiments in later sections.

2.4 The Algorithms for the HAP problem

In this section, we propose two algorithms to achieve the optimal solution for the HAP problem when the input PDFG is a simple path or a tree. For the general problem, i.e., the input is a DAG, we propose one efficient and one optimal algorithms to solve the HAP problem.

2.4.1 Definitions and Lemma

To solve the HAP problem, we use dynamic programming method traveling the graph in bottom up fashion. For the ease of explanation, we will index the nodes based on bottom up sequence. For example, Figure 2.3 (a) shows a simple path with nodes indexed in a bottom up sequence, that is, $v_1 \rightarrow v_2 \rightarrow v_3$. Figure 2.3 (b) shows a tree indexed by bottom up sequence. The sequence is: $v_1 \rightarrow v_2 \rightarrow \cdots \rightarrow v_5$.

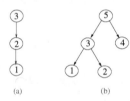

(a)   (b)

Figure 2.3. (a) A simple path with three nodes indexed in a bottom up sequence. (b) A tree with five nodes indexed in a bottom up sequence.

Given the timing constraint L, a PDFG G, and an assignment A, we first give several definitions as follows:

- **G^i**: The sub-graph rooted at node v_i, containing all the nodes reached by node v_i. In our algorithm, each step will add one node which becomes the root of its sub-graph. For example, in Figure 2.3 (b), G^3 is the tree containing nodes 1, 2, and 3.

- **$C_A(G^i)$ and $T_A(G^i)$**: The total cost and total execution time of G^i under the assignment A. In our algorithm, each step will achieve the minimum total cost of G^i with computed confidence probabilities under various timing constraints.

- In our algorithm, table $D_{i,j}$ will be built. Each entry of table $D_{i,j}$ will store a linked list of (Probability, Cost) pairs sorted by probability in ascending order. Here we define the **(Probability, Cost) pair ($P_{i,j}$, $C_{i,j}$)** as follows: $C_{i,j}$ is the minimum cost of $C_A(G^i)$ computed by all assignments A satisfying $T_A(G^i) \leq j$ with probability that is greater than or equal to $P_{i,j}$.

In every step in our algorithm, one more node will be included for consideration. The information of this node is stored in local table $B_{i,j}$, which is similar to table $D_{i,j}$. A local table store only data of probabilities and consumptions of a node itself. Table $B_{i,j}$ is the local table only storing the information of node v_i. In more detail, $B_{i,j}$ is a local table of linked lists that store pair $(p_{i,j}, c_{i,j})$ sorted by $p_{i,j}$ in an ascending order; $c_{i,j}$ is the cost only for node v_i at time j, and $p_{i,j}$ is the corresponding probability. The building procedures of $B_{i,j}$ are as follows. First, sort the execution time variations in an ascending order. Then, accumulate the probabilities of same type. Finally, let $L_{i,j}$ be the linked list in each entry of $B_{i,j}$, insert $L_{i,j}$ into $L_{i,j+1}$ while redundant pairs are canceled out. For example, node 0 in Figure 2.1(b) has the following (T: P, C) pairs: (1: 0.9, 10), (3: 0.1, 10) for type R_1, and (2: 0.7, 4), (4: 0.3, 4) for type R_2. After sorting and accumulating, we get (1: 0.9, 10), (2: 0.7, 4), (3: 1.0, 10), and (4: 1.0, 4). We obtain Table 2.5 after the insertion.

Table 2.5. An example of local table, $B_{0,j}$

$Time$	1	2	3	4
(P_i, C_i)	(0.9, 10)	(0.7, 4)	(0.7, 4)	(1.0, 4)
		(0.9, 10)	(1.0, 10)	

We introduce the **operator** "$\oplus$" in this chapter. For two (Probability, Cost) pairs H_1 and H_2, if H_1 is ($P_{i,j}^1$, $C_{i,j}^1$), and H_2 is ($P_{i,j}^2$, $C_{i,j}^2$), then, after the $\oplus$ operation between H_1 and H_2, we get pair (P', C'), where $P' = P_{i,j}^1 * P_{i,j}^2$ and $C' = C_{i,j}^1 + C_{i,j}^2$. We denote this operation as "$H_1 \oplus H_2$". This is the key operation of our algorithms. The meaning is that when two task nodes add together, the total cost is computed by adding the costs of all nodes together and the probability corresponding to the total cost is computed by multiplying the probabilities of all nodes based on the basic properties of probability and cost of a PDFG. For two independent events A and B, $P(A \cup B) = P(A) * P(B)$, and $C(A \cup B) = C(A) + C(B)$.

In our algorithm, $D_{i,j}$ is the table in which each entry has a linked list that store pair $(P_{i,j}, C_{i,j})$. Here, i represents a node number, and j represents time. For example, a linked list can be $(0.1, 2) \rightarrow (0.3, 3) \rightarrow (0.8, 6) \rightarrow (1.0, 12)$. Usually, there are redundant pairs in a linked list. We give the redundant-pair removal algorithm in Algorithm 2.4.1, which is a base for the later algorithms.

Algorithm 2.4.1 redundant-pair removal algorithm

Input: A list of $(P_{i,j}^k, C_{i,j}^k)$
Output: A redundant-pair free list
 1: Sort the list by $P_{i,j}$ in an ascending order such that $P_{i,j}^k \leq P_{i,j}^{k+1}$.
 2: From the beginning to the end of the list,
 3: **for** each two neighboring pairs $(P_{i,j}^k, C_{i,j}^k)$ and $(P_{i,j}^{k+1}, C_{i,j}^{k+1})$ **do**
 4:     **if** $P_{i,j}^k = P_{i,j}^{k+1}$ **then**
 5:         **if** $C_{i,j}^k \geq C_{i,j}^{k+1}$ **then**
 6:             cancel the pair $P_{i,j}^k, C_{i,j}^k$
 7:         **else**
 8:             cancel the pair $P_{i,j}^{k+1}, C_{i,j}^{k+1}$
 9:         **end if**
10:     **else**
11:         **if** $C_{i,j}^k \geq C_{i,j}^{k+1}$ **then**
12:             cancel the pair $(P_{i,j}^k, C_{i,j}^k)$
13:         **end if**
14:     **end if**
15: **end for**

For example, we have a list with pairs $(0.1, 2) \rightarrow (0.3, 3) \rightarrow (0.5, 3) \rightarrow (0.3, 4)$, we remove the redundant-pair as follows: First, sort the list according $P_{i,j}$ in an ascending order. This list becomes to $(0.1, 2) \rightarrow (0.3, 3) \rightarrow (0.3, 4) \rightarrow (0.5, 3)$. Second, cancel redundant pairs. Comparing $(0.1, 2)$ and $(0.3, 3)$, we keep both. For the two pairs $(0.3, 3)$ and $(0.3, 4)$, we cancel pair $(0.3, 4)$ since the cost 4 is bigger than 3 in pair $(0.3, 3)$. Comparing $(0.3, 3)$ and $(0.5, 3)$, we cancel $(0.3, 3)$ since $0.3 < 0.5$ while $3 \geq 3$. The probability 0.3 is already covered by probability 0.5 while the costs are same. There is no information lost in redundant-pair removal.

Using algorithm 2.4.1, we can cancel many redundant-pair $(P_{i,j}, C_{i,j})$ whenever we find conflicting pairs in a list during a computation. After the $\oplus$ operation and redundant pair removal, the list of ($P_{i,j}, C_{i,j}$) has the following properties:

Lemma 2.4.1. *For any $(P_{i,j}^1, C_{i,j}^1)$ and $(P_{i,j}^2, C_{i,j}^2)$ in the same list:*

1. *$P_{i,j}^1 \neq P_{i,j}^2$ and $C_{i,j}^1 \neq C_{i,j}^2$.*

2. *$P_{i,j}^1 < P_{i,j}^2$ if and only if $C_{i,j}^1 < C_{i,j}^2$.*

For two linked lists L_1 and L_2, the operation "$\mathbf{L_1} \oplus \mathbf{L_2}$" is implemented as follows: First, implement $\oplus$ operation on all possible combinations of two pairs from different linked lists. Then insert the new pairs into a new linked list and remove redundant pairs using algorithm 2.4.1.

2.4.2   An Optimal Algorithm for Simple Path

An optimal algorithm, *Path_Assign*, is proposed in the following. It can give the optimal solution for the HAP problem when the given PDFG is a simple path. For a simple path, we can use either bottom up or top down approach. Without loss of generality, we use bottom up approach, that is, starting from child to parent. Assume the node sequence of the simple path in the HAP problem is $v_1 \rightarrow v_2 \rightarrow \cdots \rightarrow v_N$, indexed in a bottom up fashion. For example, in Figure 2.3 (a), indexed in a bottom up fashion, the node sequence of the simple path is $v_1 \rightarrow v_2 \rightarrow v_3$.

The Path_Assign Algorithm

Algorithm 2.4.2 optimal algorithm for the HAP problem when the input is a simple path (*Path_Assign*)

Input: M different types of FUs, a simple path, and the timing constraint L.
Output: An optimal assignment for the simple path

1. Build a local table $B_{i,j}$ for each node of PDFG.
2.   1: let $D_{1,j} = B_{1,j}$
     2: **for** each node $v_i, i > 1$ **do**
     3:     **for** each time j **do**
     4:         **for** each time k in $B_{i,k}$ **do**
     5:             **if** $D_{i-1,j-k}! = NULL$ **then**
     6:                 $D_{i,j} = D_{i-1,j-k} \oplus B_{i,k}$
     7:             **else**
     8:                 continue
     9:             **end if**
     10:        **end for**
     11:        insert $D_{i,j-1}$ to $D_{i,j}$ and remove redundant pairs using algorithm 2.4.1.
     12:    **end for**
     13: **end for**

3. return $D_{N,j}$

In algorithm *Path_Assign*, first build a local table $B_{i,j}$ for each node. Next, in step 2 of the algorithm, when $i = 1$, there is only one node. We set the initial value, and let $D_{1,j} = B_{1,j}$. Then using dynamic programming method, build the table $D_{i,j}$. For each node v_i under each time j, we try all the times k ($1 \leq k \leq j$) in table $B_{i,j}$. We use "$\oplus$" on the two tables $B_{i,k}$ and $D_{i-1,j-k}$. Since $k + (j - k) = j$, the total time of nodes from v_1 to v_i is j. The "$\oplus$" operation add the costs of two tables together and multiply the probabilities of two tables with each other. Finally, we use algorithm 2.4.1 to cancel the conflicting (Probability, Cost) pairs. The new cost in each pair obtained in table $D_{i,j}$ is the cost of current node v_i at time k plus the cost in each pair obtained in $D_{i-1,j-k}$. Since we have used algorithm 2.4.1 canceling redundant pairs, the cost of each pair in $D_{i,j}$ is the minimum total cost for graph G^i with probability $P_{i,j}$ under timing constraint j.

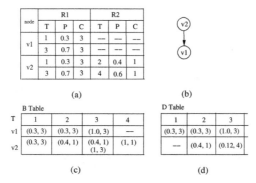

node	R1			R2		
	T	P	C	T	P	C
v1	1	0.3	3	—	—	—
	3	0.7	3	—	—	—
v2	1	0.3	3	2	0.4	1
	3	0.7	3	4	0.6	1

(a)

(b)

B Table

T	1	2	3	4
v1	(0.3, 3)	(0.3, 3)	(1.0, 3)	—
v2	(0.3, 3)	(0.4, 1)	(0.4, 1) (1, 3)	(1, 1)

(c)

D Table

	1	2	3
	(0.3, 3)	(0.3, 3)	(1.0, 3)
	—	(0.4, 1)	(0.12, 4)

(d)

Figure 2.4. (a) Initial parameters. (b) A PDFG. (c) The corresponding B table. (d) Part of corresponding D table.

For example, for the PDFG shown in Figure 2.4 (b), the initial parameters are shown in (a). We compute corresponding B table of node v_1 and v_2. For node v_2, after sorting and accumulating, we get (T: P, C) pairs: (1: 0.3, 3), (2: 0.4, 1), (3: 0.4, 1), (3: 1, 3), and (4: 1.0, 1). The results are shown in Figure 2.4 (c). Figure 2.4 (d) shows the corresponding $D_{i,j}$ table. For instance, computing $D_{2,3}$ entry, for $v_1 \rightarrow v_2$ path and buffer size $j = 3$. Buffer size $j = t(v_1) + t(v_2) = 3$. Using algorithm $Path\_Assign$, we have two cases. Case 1: $3 = 2 + 1$. Then $D_{2,3} = D_{1,2} \oplus B_{2,1}$, and $(0.3, 3) \oplus (0.4, 1) = (0.12, 4)$. Hence we get pair $(0.12, 4)$. Case 2: $3 = 1 + 2$. Then $D_{2,3} = D_{1,1} \oplus B_{2,2}$. And $(0.3, 3) \oplus (0.3, 3) = (0.09, 6)$. So we get pair $(0.09, 6)$. Since $(0.09, 6)$ is inferior to $(0.12, 4)$, it can be removed. Hence, we fill $(0.12, 4)$ into $D_{2,3}$ entry.

The cost in $D_{N,j}$ is the minimum total cost with computed confidence probability under timing constraint j. Given the timing constraint L, the minimum total cost for the graph G is the cost in $D_{N,L}$. In the following, we show Theorem 2.4.1 about this.

Theorem 2.4.1. *For each pair* $(P_{i,j},\ C_{i,j})$ *in* $D_{i,j}$ ($1 \leq i \leq N$) *obtained by algorithm* Path_Assign, $C_{i,j}$ *is the minimum total cost for graph* G^i *with confidence probability* $P_{i,j}$ *under timing constraint* j.

Proof. By induction. **Basic Step:** When $i = 1$, there is only one node and $D_{1,j} = B_{1,j}$. Thus, when $i = 1$, Theorem 2.4.1 is true. **Induction Step:** We need to show that for $i \geq 1$, if for each pair $(P_{i,j},\ C_{i,j})$ in $D_{i,j}$, $C_{i,j}$ is the minimum total cost for graph G^i with confidence probability $P_{i,j}$ under timing constraint j, then for each pair $(P_{i+1,j},\ C_{i+1,j})$ in $D_{i+1,j}$, $C_{i+1,j}$ is the minimum total cost for graph G^{i+1} with confidence probability $P_{i+1,j}$ under timing constraint j. In step 2 of the algorithm, since $j = k + (j - k)$ for each k in $B_{i+1,j}$, we try all the possibilities to obtain j. Then we use $\oplus$ operator to add the costs of two tables and multiply the probabilities of two tables. Finally, we use algorithm 2.4.1 to cancel the conflicting (Probability, Cost) pairs. The new cost in each pair obtained in table $D_{i+1,j}$ is the cost of current node $i + 1$ at time k plus the cost in each pair obtained in $D_{i,j-k}$. Since we have used algorithm 2.4.1 to cancel redundant pairs, the cost of each pair in $D_{i+1,j}$ is the minimum total cost for graph G^{i+1} with confidence probability $P_{i+1,j}$ under timing constraint j. Thus, Theorem 2.4.1 is true for any i ($1 \leq i \leq N$). $\qquad\square$

From Theorem 2.4.1, we know $D_{N,L}$ records the minimum total cost of the whole path with corresponding confidence probabilities under the timing constraint L. We can record the corresponding FU type assignment of each node when computing the minimum total cost in step 2 in the algorithm *Path_Assign*. Using these information, we can get an optimal assignment by tracing how to reach $D_{N,L}$.

It takes $O(M * K)$ to compute one value of $D_{i,j}$, where M is the maximum number of FU types, and K is the maximum number of execution time variations for each node. Thus, the complexity of the algorithm *Path_Assign* is $O(|V| * L * M * K)$, where $|V|$ is the number of nodes and L is the given timing constraint. Usually, the execution time of each node is upper bounded by a constant. So L equals $O(|V|^c)$ (c is a constant). In this case, *Path_Assign* is polynomial.

2.4.3 An Optimal Algorithm For Tree

In this section, we propose an optimal algorithm, *Tree_Assign*, to produce the optimal solution to the HAP problem when the input PDFG is a tree.

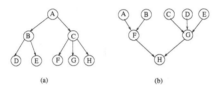

(a)   (b)

Figure 2.5. (a)Multi-child tree (b)Multi-parent tree.

Define a *root* node to be a node without any parent and a *leaf* node to be a node without any child. In multi-child case, we use bottom up approach. For example, both $\{D, E, B\}$ and $\{E, D, B\}$ are bottom up sequences for the given tree in Figure 2.5(a). Here, sequences do not matter as long as bottom up is followed, since bottom up is used to guarantee that, when we begin to process a node, the processing of all of its child nodes has already been finished in the algorithm. So, there is no difference between $\{D, E, B\}$ and $\{E, D, B\}$ in terms of our algorithm. The reason we use bottom up approach in Figure 2.5(a) is that we can merge two children first as one node, then we get the simple path. Hence we can get the optimal assignment. If using top down approach, we have two paths. It is possible there are two optimal assignments for the root node and other nodes are common to both paths. For example, in Figure 2.5(a), if we use top down approach to execute the dynamic program, the best assignments for routes $A \rightarrow B \rightarrow D$ and $A \rightarrow B \rightarrow E$

may have different assignments at node A. Hence we should use bottom up approach. Figure 2.5(b) is a multi-parent tree, we should use top down approach for the same reason. The pseudo polynomial algorithm for trees is shown in Algorithm 2.4.3.

Algorithm 2.4.3 optimal algorithm for the HAP problem when the input is a tree (*Tree_Assign*)

Input: M different types of FUs, a tree, and the timing constraint L.
Output: An optimal assignment for the tree

1.   **if** the tree is multi-child type **then**
        use bottom up approach (leaves to root)
    else
        use top down approach (roots to leaf)
    end if

2. For bottom up approach, use the following algorithm.
    For top down approach, just reverse the tree.
    $|V| \leftarrow N$, where $|V|$ is the number of nodes.
    Assume the sequence of the tree is $v_1 \rightarrow v_2 \rightarrow \cdots \rightarrow v_N$, in bottom up fashion.
    Let $D_{1,j} = B_{1,j}$.
    Assume $D'_{i,j}$ is the table that stored minimum total cost with computed confidence probabilities under the timing constraint j for the subtree rooted on v_i except v_i.
    Nodes $v_{i_1}, v_{i_2}, \cdots, v_{i_w}$ are all child nodes of node v_i.
    w is the number of child nodes of node v_i, then

$$D'_{i,j} = \begin{cases} (0,0) & \text{if } w = 0 \\ D_{i_1,j} & \text{if } w = 1 \\ D_{i_1,j} \oplus D_{i_2,j} \oplus \cdots \oplus D_{i_w,j} & \text{if } w \geq 1 \end{cases} \qquad (2.1)$$

    Then, for each k in $B_{i,k}$.

$$D_{i,j} = D'_{i,j-k} \oplus B_{i,k} \qquad (2.2)$$

3. return $D_{N,j}$

In *Tree_Assign* algorithm, first, we decide which approach to use. If the tree is multi-child type, then use bottom up approach (leaves to root). Otherwise, we use top down approach (roots to leaf). Without loss of generality, we use bottom up approach. when $w = 1$, node v_i has only one child v_{i_1}. By using *Path_Assign* algorithm, we get $D_{i,j} = (P_{i,j}, C_{i,j})$, where $P_{i,j} = P_{i_1,j-k} * P_{i,k}$, and $C_{i,j} = C_{i_1,j-k} + C_{i,k}$, by using the operator "⊕". If $w \geq 1$, node v_i has multiple children. We merge all the children of node v_i into one pseudo child. Then we can use *Path_Assign* algorithm to get final solution. The merging procedures are as follows. At the same time j, sum up the costs of all children and multiply the probabilities of all children. For instance, if node v_i has two child nodes v_{i_1} and v_{i_2}, then $D'_{i,j} = (P'_{i,j}, C'_{i,j})$, where $P'_{i,j} = P_{i_1,j} * P_{i_2,j}$, and $C'_{i,j} = C_{i_1,j} + C_{i_2,j}$. After merging all children into one pseudo child, we can continue implement *Path_Assign* algorithm to get the final solution for the tree.

According to the bottom up approach in multi-child tree (for top down approach, just reverse the tree), the execution of $D_{i-1,j}$ for each child node of v_i has been finished before executing $D_{i,j}$. From equation (2.1), $D'_{i,j}$ gets the summation of the minimum total cost of all child nodes of v_i because they can be executed simultaneously within time j. From equation (2.2), the minimum total cost is selected from all possible costs caused by adding v_i into the graph G^i. Therefore, for each pair $(P_{i,j}, C_{i,j})$ in $D_{i,j}$, $C_{i,j}$ is the minimum total cost for the graph G^i with confidence probability $P_{i,j}$ under timing constraint j.

Algorithm *Tree_Assign* gives the optimal solution when the given PDFG is a tree. We have Theorem 2.4.2 about this.

Theorem 2.4.2. *For each pair $(P_{i,j}, C_{i,j})$ in $D_{i,j}$ $(1 \leq i \leq N)$ obtained by algorithm Tree_Assign, $C_{i,j}$ is the minimum total cost for the graph G^i with confidence probability $P_{i,j}$ under timing constraint j.*

Proof. By induction. **Basic Step:** When $i = 1$, There is only one node and $D_{1,j} = B_{1,j}$. Thus, when $i = 1$, Theorem 2.4.2 is true. **Induction Step:** We need to show that for $i \geq 1$, if for each pair $(P_{i,j}, C_{i,j})$ in $D_{i,j}$, $C_{i,j}$ is the minimum total cost for the graph G^i with confidence probability $P_{i,j}$ under timing constraint j, then for each pair $(P_{i+1,j}, C_{i+1,j})$ in $D_{i+1,j}$, $C_{i+1,j}$ is the total system cost for the graph G^{i+1} with confidence probability $P_{i+1,j}$ under timing constraint j. According to the bottom up approach in multi-child tree (for top down approach, just reverse the tree), the execution of $D_{i,j}$ for each child node of v_{i+1} has been finished before executing $D'_{i+1,j}$. From equation (2.1), $D'_{i+1,j}$ gets the summation of the minimum total cost of all child nodes of v_{i+1} because they can be executed simultaneously within time j. From equation (2.2), the minimum total cost is selected from all possible costs caused by adding v_{i+1} into the graph G^{i+1}. So for each pair $(P_{i+1,j}, C_{i+1,j})$ in $D_{i+1,j}$, $C_{i+1,j}$ is the minimum total cost for the graph G^{i+1} with confidence probability $P_{i+1,j}$ under timing constraint j. Therefore, Theorem 2.4.2 is true for any i $(1 \leq i \leq N)$. □

The complexity of algorithm *Tree_Assign* is $O(|V| * L * M * K)$, where $|V|$ is the number of nodes, L is the given timing constraint, M is the maximum number of FU types for each node, and K is the maximum number of execution time variation for each node. When L equals $O(|V|^c)$ (c is a constant) which is the general case in practice, algorithm *Tree_Assign* is polynomial.

2.4.4 A Near-Optimal Heuristic Algorithm for DAG

In this and next subsections, we propose two algorithms to solve the general case of the HAP problem, i.e., the given input is a *directed acyclic graph* (DAG). Since the general problem of the HAP problem is NP-complete problem, one of the two algorithms we proposed here is near-optimal heuristic algorithm, and the other is an optimal one. In many cases, the near-optimal heuristic algorithm will give us the same results as the results of the optimal algorithm. The optimal algorithm is suitable for the cases when the given PDFG has small number of multi-parent and multi-child nodes.

We give an example of DAG. The input PDFG is shown in Figure 2.6(a), which has five nodes. The times, costs, and probabilities of each node is shown in Figure 2.6(b). Node 4 is a multi-child node, which has three children: 0, 2, and 3. Node 0 is a multi-parent

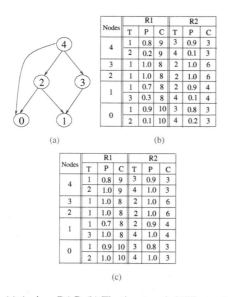

(a)

Nodes	R1			R2		
	T	P	C	T	P	C
4	1	0.8	9	3	0.9	3
	2	0.2	9	4	0.1	3
3	1	1.0	8	2	1.0	6
2	1	1.0	8	2	1.0	6
1	1	0.7	8	2	0.9	4
	3	0.3	8	4	0.1	4
0	1	0.9	10	3	0.8	3
	2	0.1	10	4	0.2	3

(b)

Nodes	R1			R2		
	T	P	C	T	P	C
4	1	0.8	9	3	0.9	3
	2	1.0	9	4	1.0	3
3	1	1.0	8	2	1.0	6
2	1	1.0	8	2	1.0	6
1	1	0.7	8	2	0.9	4
	3	1.0	8	4	1.0	4
0	1	0.9	10	3	0.8	3
	2	1.0	10	4	1.0	3

(c)

Figure 2.6. (a) A given DAG. (b) The times, probabilities, and costs of its nodes for different FU types. (c) The time cumulative distribution functions (CDFs) and costs of its nodes for different FU types.

node, and has two parents: 2 and 4. Figure 2.6(c) shows the time cumulative distribution functions (CDFs) and costs of each node for different FU types.

We give the near-optimal heuristic algorithm (*DAG_Heu*) for the HAP problem when the given PDFG is a DAG, which is shown in Algorithm 2.4.4.

In algorithm *DAG_Heu*, if using bottom up approach, for each sequence node, use the simple path algorithm to get the dynamic table of parent node. If using top down approach, reverse the sequence and use the same algorithm. For example, in Figure 2.6(a), there are two multi-child nodes: 2 and 4, and there are also two multi-parent nodes, that is, node 0 and 1. Hence, we can use either approach.

Algorithm 2.4.4 *DAG_Heu* Algorithm

Input: M different types of FUs, a DAG, and the timing constraint L.
Output: A near-optimal heuristic assignment for the DAG

1. $SEQ \leftarrow$ Sequence obtained by topological sorting all the nodes.

2. $t_{mp} \leftarrow$ the number of multi-parent nodes;
   $t_{mc} \leftarrow$ the number of multi-child nodes;
   If $t_{mp} < t_{mc}$, use bottom up approach;
   else, use top down approach.

3. If bottom up approach, use the following algorithm;
   If top down approach, just reverse the sequence.
   $|V| \leftarrow N$, where $|V|$ is the number of nodes.

4. $SEQ \leftarrow \{v_1 \rightarrow v_2 \rightarrow \cdots \rightarrow v_N\}$, in bottom up fashion;
   $D_{1,j} \leftarrow B_{1,j}$;
   $D'_{i,j} \leftarrow$ the table that stored MIN(C) with $Prob.(T \leq j) \geq P$ for the sub-graph rooted on v_i
   except v_i;
   $v_{i_1}, v_{i_2}, \cdots, v_{i_w} \leftarrow$ all child nodes of node v_i;
   $w \leftarrow$ the number of child nodes of node v_i.

$$D'_{i,j} = \begin{cases} (0,0) & \text{if } w = 0 \\ D_{i_1,j} & \text{if } w = 1 \\ D_{i_1,j} \oplus \cdots \oplus D_{i_w,j} & \text{if } w > 1 \end{cases} \tag{2.3}$$

5. Computing $D_{i_1,j} \oplus D_{i_2,j}$:
   $G' \leftarrow$ the union of all nodes in the graphs rooted at nodes v_{i_1} and v_{i_2};
   Travel all the graphs rooted at nodes v_{i_1} and v_{i_2};
   If a node is a common node, then use a selection function to choose the FU type of a node.

6. For each k in $B_{i,k}$.

$$D_{i,j} = D'_{i,j-k} \oplus B_{i,k} \tag{2.4}$$

7. Then use the algorithm 2.4.1 to remove redundant pairs;
   $D_{N,j} \leftarrow$ a table of MIN(C) with $Prob.(T \leq j) \geq P$;
   Output $D_{N,L}$.

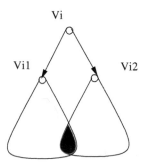

Figure 2.7. Problem of common nodes.

In algorithm *DAG_Heu*, we also have to solve the problem of common nodes, that is, one node appears in two or more graphs that rooted by the child nodes of node v_i. In equation (2.3), even if there are common nodes, we must not count the same node twice. That is, the cost is just added once, and the probability is multiplied once. For example, in Figure 2.7, the two children of node v_i are v_{i_1} and v_{i_2}. The two subgraphs that rooted at nodes v_{i_1} and v_{i_2} have common areas (the shaded area in Figure 2.7). The nodes in this common area are called *common nodes*. When we compute the cost and probability of node v_i, we can only count these common nodes once. If a common node has conflicting FU type selection, then we need to define a selection function to decide which FU type should be chosen for the common node. For example, we can select the FU type that has a smaller execution time as the FU type of a common node.

Due to the problem of common nodes, algorithm *DAG_Heu* is not an optimal algorithm. The reason is that an assignment conflict for a common node maybe exist, while algorithm *DAG_Heu* cannot solve this problem. For example, in Figure 2.6, we focus on the subgraph composed of nodes 1, 2, 3, and 4. Using bottom up approach, there will be two paths from node 1 to node 4. Path a is $1 \rightarrow 2 \rightarrow 4$, and path b is $1 \rightarrow 3 \rightarrow 4$. Hence, node 1 is a common node for both paths while node 4 is the root. It is possible, under a timing constraint, the best assignment for path a gives node 1 assignment as FU type 1, while the best assignment for path b gives node 1 assignment as FU type 2. This kind of assignment conflicting can not be solved by algorithm *DAG_Heu*. Hence, *DAG_Heu* is not an optimal algorithm, although it is very efficient in practice.

From algorithm *DAG_Heu*, we know $D_{N,L}$ records the minimum total cost of the whole path within the timing constraint L. We can record the corresponding FU type assignment of each node when computing the minimum system cost in the algorithm *DAG_heu*. Using these information, we can get an optimal assignment by tracing how to reach $D_{N,L}$.

It takes at most $O(|V|)$ to compute common nodes for each node in the algorithm *DAG_heu*, where $|V|$ is the number of nodes. Thus, the complexity of the algorithm *DAG_heu* is $O(|V|^2 * L * M * K)$, where L is the given timing constraint, M is the maximum number of FU types for each node, and K is the maximum number of execution time variation for each node. Usually, the execution time of each node is upper bounded by a constant. Then L equals $O(|V|^c)$ (c is a constant). In this case, *DAG_heu* is polynomial.

2.4.5 An Optimal Algorithm for DAG

In this subsection, we give the optimal algorithm (*DAG_Opt*) for the HAP problem when the given PDFG is a DAG. In *DAG_Opt*, we exhaust all the possible assignments of multi-parent or multi-child nodes. Without loss of generality, assume we using bottom up approach. If the total number of nodes with multi-parent is t, and there are maximum K variations for the execution times of all nodes, then we will give each of these t nodes a fixed assignment. We will exhaust all of the K^t possible fixed assignments by algorithm *DAG_Heu* without using the selection function.

Algorithm 2.4.5 *DAG_Opt* Algorithm

Input: M different types of FUs, a DAG, and the timing constraint L.
Output: An optimal assignment for the DAG

1. $SEQ \leftarrow$ Sequence obtained by topological sorting all the nodes.

2. $t_{mp} \leftarrow$ the number of multi-parent nodes;
   $t_{mc} \leftarrow$ the number of multi-child nodes;
   If $t_{mp} < t_{mc}$, use bottom up approach;
   else, use top down approach.

3. If bottom up approach, use the following algorithm;
   If top down approach, just reverse the sequence.
   $|V| \leftarrow N$, where $|V|$ is the number of nodes.

4. If the total number of nodes with multi-parent is t, and there are maximum K variations for the execution data loads of all nodes, then we will give each of these t nodes a fixed assignment.

5. For each of the K^t possible fixed assignments,
   $SEQ \leftarrow \{v_1 \rightarrow v_2 \rightarrow \cdots \rightarrow v_N\}$, in bottom up fashion;
   $D_{1,j} \leftarrow B_{1,j}$;
   $D'_{i,j} \leftarrow$ the table that stored MIN(C) with $Prob.(T \leq j) \geq P$ for the sub-graph rooted on v_i except v_i;
   $v_{i_1}, v_{i_2}, \cdots, v_{i_w} \leftarrow$ all child nodes of node v_i;
   $w \leftarrow$ the number of child nodes of node v_i.

$$D'_{i,j} = \begin{cases} (0,0) & \text{if } w = 0 \\ D_{i_1,j} & \text{if } w = 1 \\ D_{i_1,j} \oplus \cdots \oplus D_{i_w,j} & \text{if } w > 1 \end{cases} \qquad (2.5)$$

6. Computing $D_{i_1,j} \oplus D_{i_2,j}$:
   $G' \leftarrow$ the union of all nodes in the graphs rooted at nodes v_{i_1} and v_{i_2};
   Travel all the graphs rooted at nodes v_{i_1} and v_{i_2};

7. For each k in $B_{i,k}$.

$$D_{i,j} = D'_{i,j-k} \oplus B_{i,k} \qquad (2.6)$$

8. For each possible fixed assignment, we get a $D_{N,j}$. Merge the (P, C) pairs in all the possible $D_{N,j}$ together, and sort them in ascending sequence according P.

9. Then use the Algorithm 2.4.1 to remove redundant pairs;
   $D_{N,j} \leftarrow$ a table of MIN(C) with $Prob.(T \leq j) \geq P$;
   Output $D_{N,L}$.

Algorithm *DAG_Opt* gives the optimal solution when the given PDFG is a DAG, which is shown in Algorithm 2.4.5. In the following, we give the Theorem 2.4.3 and Theorem 2.4.4 about this.

Theorem 2.4.3. *In each possible fixed assignment, for each pair $(P_{i,j}, C_{i,j})$ in $D_{i,j}$ ($1 \le i \le N$) obtained by algorithm* DAG_Opt, $C_{i,j}$ *is the minimum total cost for the graph G^i with confidence probability $P_{i,j}$ under timing constraint j.*

Proof. By induction.

Basic Step: When $i = 1$, There is only one node and $D_{1,j} = B_{1,j}$. Thus, when $i = 1$, Theorem 2.4.3 is true. **Induction Step:** We need to show that for $i \ge 1$, if for each pair $(P_{i,j}, C_{i,j})$ in $D_{i,j}$, $C_{i,j}$ is the minimum total cost of the graph G^i, then for each pair $(P_{i+1,j}, C_{i+1,j})$ in $D_{i+1,j}$, $C_{i+1,j}$ is the total cost of the graph G^{i+1} with confidence probability $P_{i+1,j}$ under timing constraint j. According to the bottom up approach (for top down approach, just reverse the sequence), the execution of $D_{i,j}$ for each child node of v_{i+1} has been finished before executing $D_{i+1,j}$. From equation (2.5), $D'_{i+1,j}$ gets the summation of the minimum total cost of all child nodes of v_{i+1} because they can be executed simultaneously within time j. We avoid the repeat counting of the common nodes. Hence, each nodes in the graph rooted by node v_{i+1} was counted only once. From equation (2.6), the minimum total cost is selected from all possible costs caused by adding v_{i+1} into the subgraph rooted on v_{i+1}. So for each pair $(P_{i+1,j}, C_{i+1,j})$ in $D_{i+1,j}$, $C_{i+1,j}$ is the total cost of the graph G^{i+1} with confidence probability $P_{i+1,j}$ under timing constraint j. Therefore, Theorem 2.4.3 is true for any i ($1 \le i \le N$). □

Theorem 2.4.4. *For each pair $(P_{i,j}, C_{i,j})$ in $D_{N,j}$ ($1 \le j \le L$) obtained by algorithm* DAG_Opt, $C_{i,j}$ *is the minimum total cost for the given DAG G with confidence probability $P_{i,j}$ under timing constraint j.*

Proof. According to Theorem 2.4.3, in each possible fixed assignment, for each pair $(P_{i,j}$, $C_{i,j})$ in $D_{i,j}$ we obtained, $C_{i+1,j}$ is the total cost of the graph G^{i+1} with confidence probability $P_{i+1,j}$ under timing constraint j. In step 4 of the algorithm *DAG_Opt*, we try all the possible fixed assignments, combine them together into a new row $D_{N,j}$ in dynamic table, and remove redundant pairs using the algorithm 2.4.1. Hence, for each pair $(P_{i,j}, C_{i,j})$ in $D_{N,j}$ $(1 \leq j \leq L)$ obtained by algorithm *DAG_Opt*, $C_{i,j}$ is the minimum total cost for the given DAG G with confidence probability $P_{i,j}$ under timing constraint j. $\square$

In algorithm *DAG_Opt*, there are K^t loops and each loop needs $O(|V|^2 * L * M * K)$ running time. The complexity of *Algorithm DAG_Opt* is $O(K^t * |V|^2 * L * M * K)$, where t is the total number of nodes with multi-parent (or multi-child) in bottom up approach (or top down approach), $|V|$ is the number of nodes, L is the given timing constraint, M is the maximum number of FU types for each node, and K is the maximum number of execution time variation for each node. Algorithm *DAG_Opt* is exponential, hence it can not be applied to a graph with large amounts of multi-parent and multi-child nodes.

For Figure 2.6, the minimum total costs with computed confidence probabilities under the timing constraint are shown in Table 2.6. The entries with probability equal to 1 (see the entries in boldface) actually give the results to the hard HA problem which shows the worst-case scenario of the HAP problem. In this example, the algorithm *DAG_Heu* gives the same results as those of the algorithm *DAG_Opt*. Actually, experiments shown that although algorithm *DAG_Heu* is only near-optimal, it can give same results as those given by the optimal algorithm in most cases.

2.5 Experiments

This section presents the experimental results of our algorithms. We conduct experiments on a set of benchmarks including 4-stage lattice filter, 8-stage lattice filter, voltera filter,

Table 2.6. Minimum total costs with computed confidence probabilities under various timing constraints for a DAG.

T	(P , C)	(P , C)	(P , C)	(P , C)	(P , C)	(P , C)
3	0.50, 43					
4	0.72, 39					
5	0.58, 30	0.72, 35	0.90, 39			
6	0.58, 28	0.72, 30	0.80, 32	0.81, 33	0.90, 35	**1.00, 43**
7	0.65, 24	0.80, 28	0.81, 29	0.90, 30	**1.00, 32**	
8	0.65, 22	0.81, 24	0.90, 26	**1.00, 28**		
9	0.90, 22	**1.00, 26**				
10	**1.00, 22**					

differential equation solver, RLS-languerre lattice filter, and elliptic filter. Among them, the PDFG for first three filters are trees and those for the others are DAGs. The basic information about the benchmarks is shown in Table 2.7, in which a multi-parent node is a node with more than one parent and a multi-child node is a node with more than one child. Three different FU types, R_1, R_2, and R_3, are used in the system, in which a FU with type R_1 is the quickest with the highest cost and a FU with type R_3 is the slowest with the lowest cost. The distribution of execution times of each node is Gaussian. For each benchmark, the first timing constraint we use is the minimum execution time. The experiments are performed on a Dell PC with a P4 2.1 G processor running Red Hat Linux 7.3.

Table 2.7. The basic information for the Benchmarks

Benchmarks	PDFG	# of nodes	# of multi-parent	# of multi-child
voltera	Tree	27		
4-lat-iir	Tree	26		
8-lat-iir	Tree	42		
Diff. Eq.	DAG	11	3	1
RLS-lagu.	DAG	19	6	3
elliptic	DAG	34	8	5

The experiments on voltera filter, 4-stage lattice filter, and 8-stage lattice filter are finished in less than one second, in which we compare our algorithms with the *hard HA ILP*

Table 2.8. The minimum total costs with computed confidence probabilities under various timing constraints for voltera filter.

Voltera Filter							
TC	0.7		0.8		0.9		1.0
	cost	%	cost	%	cost	%	cost
62	7896		×		×		×
80	7166		7169		7847		×
100	5366	31.5	5369	31.4	6047	22.8	**7827**
125	5347	31.7	5352	31.6	5843	25.3	**7820**
150	4032	43.8	4066	43.6	4747	32.8	**7169**
175	1604	66.2	2247	52.7	2947	37.9	**4747**
200	1587	66.3	1618	65.6	2318	50.7	**4704**
225	1580	46.4	1593	45.9	1647	44.1	**2947**
250	1580	31.9	1582	31.8	1604	30.8	**2318**
273	1580	4.1	1580	4.1	1580	4.1	**1647**
274	1580		1580		1580		**1580**
Ave. Redu.(%)	40.2		38.3		31.0		

Table 2.9. The minimum total costs with computed confidence probabilities under various timing constraints for 4-stage lattice filter.

4-stage Lattice IIR Filter							
TC	0.7		0.8		0.9		1.0
	cost	%	cost	%	cost	%	cost
81	3462		×		×		×
100	3452		3472		×		×
125	3452		2290		3525		×
150	1881		2257		2690		×
166	1858	47.3	2250	36.2	2290	35.0	**3525**
175	1853	46.8	1890	45.7	1890	45.7	**3481**
200	1325	50.4	1325	50.4	1462	45.3	**2672**
226	1259	15.5	1259	15.5	1450	2.7	**1490**
227	1259		1259		1259		**1259**
Ave. Redu.(%)	40.0		36.9		32.4		

Table 2.10. The minimum total costs with computed confidence probabilities under various timing constraints for 8-stage lattice filter.

8-stage Lattice IIR Filter							
TC	0.7		0.8		0.9		1.0
	cost	%	cost	%	cost	%	cost
94	4543		×		×		×
100	4499		5039		×		×
125	1870		2375		4539		×
144	1863	66.4	1863	66.4	2380	57.1	**5543**
150	1820	66.9	1849	66.3	2362	57.0	**5495**
175	795	67.4	951	61.0	1339	45.1	**2439**
200	732	43.5	732	43.5	775	40.2	**1295**
225	595	29.1	638	23.8	639	23.8	**839**
250	532	12.9	540	11.6	540	11.6	**611**
277	506	4.9	511	4.0	511	4.0	**532**
278	506		506		506		**506**
Ave. Redu.(%)	41.6		39.5		34.1		

algorithm [43]. The experimental results for these filters are shown in Table 2.8-2.10. In each table, column "TC" represents the given timing constraint. The minimum total costs obtained from different algorithms: *Tree_Assign* and the optimal *hard HA ILP algorithm* [43], are presented in each entry. Columns "1.0", "0.9", "0.8", and "0.7" represent that the confidence probability is 1.0, 0.9, 0.8, and 0.7, respectively. Algorithm *Tree_Assign* covers all the probability columns, while the optimal *hard HA ILP algorithm* [43] only include the column "1.0", which is in boldface. For example, in Table 2.8, under the timing constraint 100, the entry under "1.0" is 7827, which is the minimum total cost for the hard HA problem. The entry under "0.9" is 6047, which means we can achieve minimum total cost 6047 with confidence probability 0.9 under timing constraints. From the information provided in the structure of the linked list in each entry of the dynamic table, we are able to trace how to get the satisfied assignment.

Column "%" shows the percentage of reduction on the total cost, compared the results of algorithm with those obtained by the optimal *hard HA ILP algorithm* [43]. The average percentage reduction is shown in the last row "Ave. Redu(%)" of all Tables 2.8-

2.10. The entry with "×" means no solution available. In Table 2.8, under timing constraint 80, the optimal *hard HA ILP algorithm* [43] can not find a solution. However, we can find solution 7847 with 0.9 probability that guarantees the total execution time of the PDFG are less than or equal to the timing constraint 80.

Through the experimental results, we find that our algorithms have better performance compared with the optimal *hard HA ILP algorithm* [43]. On average, algorithm *Tree_Assign* gives a cost reduction of 32.5% with 0.9 confidence probability under timing constraints, and a cost reduction of 38.2% and 40.6% with 0.8 and 0.7 confidence probability satisfying timing constraints, respectively.

Table 2.11. The minimum total costs with computed confidence probabilities under various timing constraints for differential equation solver.

Diff. Eq. Solver							
TC	0.7		0.8		0.9		1.0
	cost	%	cost	%	cost	%	cost
35	5320		×		×		×
40	5130		5230		×		×
50	5020		5040		5100		×
60	3330	38.6	3430	36.7	3830	29.3	**5420**
70	2730	49.7	2830	47.9	3230	40.5	**5430**
80	2590	47.1	2620	46.5	3020	38.4	**4900**
90	2520	48.4	2540	48.0	3100	36.5	**4880**
100	1910	38.2	1930	37.5	2690	12.9	**3090**
117	1970	32.5	1970	32.5	1970	32.5	**2920**
118	1970		1970		1970		**1970**
Ave. Redu.(%)		42.4		41.5		31.7	

We also conduct experiments on different equation solver, RLS-Laguerre filter, and elliptic filter, which are DAGs. Different equation solver has 1 multi-child node and 3 multi-parent nodes. Using top-down approach, we exhaust all 3 possible fixed assignments. RLS-Laguerre filter has 3 multi-child nodes and 6 multi-parent nodes. Using top-down approach, we implement all $3^3 = 27$ possibilities. Elliptic filter has 5 multi-child nodes and 8 multi-parent nodes. There are total $3^5 = 243$ possibilities by top-down approach.

Table 2.12. The minimum total costs with computed confidence probabilities under various timing constraints for rls-laguerre filter.

	RLS-Laguerre filter						
TC	0.7		0.8		0.9		1.0
	cost	%	cost	%	cost	%	cost
49	7803		×		×		×
60	7790		7791		7793		×
70	7082		7087		7787		×
80	5403	30.6	5991	23.0	5993	23.0	**7780**
100	3969	48.9	4669	39.9	5380	30.8	**7769**
125	2165	59.8	2269	58.0	4664	13.6	**5390**
150	1564	66.4	2264	49.3	2864	38.6	**4667**
175	1564	66.5	1564	66.5	2264	51.5	**4664**
205	1564	30.9	1564	30.9	1564	30.9	**2264**
206	1564		1564		1564		**1564**
Ave. Redu.(%)		50.5		44.6		31.4	

The experimental results for these filters are shown in Table 2.11-2.13. Column "%" shows the percentage of reduction on system cost, compared the results for soft real-time with those for hard real-time. The average percentage reduction is shown in the last row "Ave. Redu(%)" of all these three tables. The entry with "×" means no solution available. Under timing constraint 50 in Table 2.11, there is no solution for hard real-time. However, we can find solution 5100 with 0.9 probability that guarantees the total execution time of the PDFG are less than or equal to the timing constraint 50.

The experimental results show that our algorithm can greatly reduce the total cost while have a guaranteed confidence probability satisfying the timing constraints. On average, algorithm *DAG_Opt* gives a cost reduction of 33.5% with 0.9 confidence probability under timing constraints, and a cost reduction of 45.3% and 48.9% with 0.8 and 0.7 confidence probabilities satisfying timing constraints, respectively. The experiments using *DAG_Heu* algorithm on these benchmarks are finished within several seconds and the experiments using *DAG_Opt* algorithm on these benchmarks are finished within several minutes.

Table 2.13. The minimum total costs with computed confidence probabilities under various timing constraints for elliptic filter.

TC	Elliptic Filter						
	0.7		0.8		0.9		1.0
	cost	%	cost	%	cost	%	cost
120	6025		×		×		×
130	5685		5722		×		×
140	4685		4722		7524		×
150	4661		4680		5721		×
157	2585	65.5	4681	37.6	5681	24.2	7496
160	2604	65.3	2641	64.8	5703	24.1	7511
170	1983	73.4	2571	65.5	4382	41.2	7449
180	1900	70.7	1933	70.2	2612	59.7	6482
190	1850	57.4	1872	56.9	2533	41.7	4344
200	1816	57.8	1823	57.6	1933	55.1	4301
210	1803	57.6	1807	57.5	1881	55.8	4257
220	1798	30.3	1798	30.3	1828	29.1	2579
230	1796	7.1	1796	7.1	1822	5.7	1933
231	1796		1796		1796		1796
Ave. Redu.(%)	53.9		49.7		37.4		

The experimental results are same for the experiments on these three DAG benchmarks using algorithm *DAG_Heu* or *DAG_Opt*. When the number of multi-parent and multi-child nodes is large, we can use algorithm *DAG_Heu*, which will give good results in most cases.

The advantages of our algorithms over the *hard HA ILP algorithm* [43] are summarized as follows. First, our algorithms are efficient and provide overview of all possible variations of minimum costs comparing with the the worst-case scenario generated by the *hard HA ILP algorithm* [43]. More information and choices are provided by our algorithms. Second, it is possible to greatly reduce the system total cost while have a very high confidence probability under different timing constraints. Third, given an assignment, we are able to get a minimum total cost with different confidence probabilities under each timing constraint. Finally, our algorithms are very quick and practical.

2.6 Conclusion

This chapter proposed a probability approach for addressing high-level synthesis of special purpose architectures for real-time embedded systems using heterogeneous functional units with probabilistic execution time. For the *heterogeneous assignment with probability* (HAP) problem, Algorithms, *Path_Assign* and *Tree_Assign*, were proposed to give the optimal solutions when the input graph are a simple path and a tree, respectively. Two other algorithms, one is optimal and the other is near-optimal heuristic, were proposed to solve the general problem. Experiments showed that our algorithms achieve significant energy-saving and provide more design choices to achieve minimum total cost while the timing constraint is satisfied with a guaranteed confidence probability. Our algorithms are useful for both hard and soft real-time systems.

CHAPTER 3

VOLTAGE ASSIGNMENT WITH GUARANTEED PROBABILITY SATISFYING TIMING CONSTRAINT FOR REAL-TIME MULTIPROCEESOR DSP

Dynamic Voltage Scaling (DVS) is one of the techniques used to obtain energy-saving in real-time DSP systems. In many DSP systems, some tasks contain conditional instructions that have different execution times for different inputs. Due to the uncertainties in execution time of these tasks, this chapter models each varied execution time as a probabilistic random variable and solves the *Voltage Assignment with Probability* (VAP) Problem. VAP problem involves finding a voltage level to be used for each node of a probabilistic date flow graph (PDFG) in uniprocessor and multiprocessor DSP systems. This chapter proposes two optimal algorithms, one for uniprocessor and one for multiprocessor DSP systems, to minimize the expected total energy consumption while satisfying the timing constraint with a guaranteed confidence probability. The experimental results show that our approach achieves significant energy saving than previous work. For example, our algorithm for multiprocessor achieves an average improvement of 56.1% on total energy-saving with 0.80 probability satisfying timing constraint.

3.1 Introduction

The increasingly ubiquitous DSP systems pose great challenges different from those faced by general-purpose computers. DSP systems are more application specific and more constrained in terms of power, timing, and other resources. Energy-saving is a critical issue and performance metric in DSP systems design due to wide use of portable devices, espe-

71

cially those powered by batteries [16, 17, 28, 50, 65, 101]. The systems become more and more complicate and some tasks may not have fixed execution time. Such tasks usually contain conditional instructions and/or operations that could have different execution times for different inputs [36–38, 92, 111]. It is possible to obtain the execution time distribution for each task by sampling and knowing detailed timing information about the system or by profiling the target hardware [91]. Also some multimedia applications, such as image, audio, and video data streams, often tolerate occasional deadline misses without being noticed by human visual and auditory systems. For example, in packet audio applications, loss rates between 1% - 10% can be tolerated [11].

Prior design space exploration methods for hardware/software codesign of DSP systems [42, 43, 80, 95] guarantee no deadline missing by considering worst-case execution time of each task. Many design methods have been developed based on worst-case execution time to meet the timing constraints without any deadline misses. These methods are pessimistic and are suitable for developing systems in a hard real-time environment, where any deadline miss will be catastrophic. However, there are also many soft real-time systems, such as heterogeneous systems, which can tolerate occasional violations of timing constraints. The above pessimistic design methods can't take advantage of this feature and will often lead to over-designed systems that deliver higher performance than necessary at the cost of expensive hardware, higher energy consumption, and other system resources.

There are several papers on the probabilistic timing performance estimation for soft real-time systems design [36–38, 45, 68, 91, 92, 111]. The general assumption is that each task's execution time can be described by a discrete probability density function that can be obtained by applying path analysis and system utilization analysis techniques. Hu et al. [111] propose a state-based probability metric to evaluate the overall probabilistic timing performance of the entire task set. However, their evaluation method becomes very time consuming when task has many different execution time variations. Hua et al. [37, 38] propose the concept of *probabilistic design* where they design the system to meet the timing constraints of periodic applications statistically. But their algorithm is not optimal and only suitable to uniprocessor executing tasks according to a fixed order, that is, a simple path. In this chapter, we will propose an optimal algorithm for the uniprocessor situation. Also, we will give an optimal algorithm for multiprocessor executing tasks according to an executing order in DAG (Direct Acyclic Graph).

Low power and low energy consumptions are extremely important for real-time DSP systems. Dynamic voltage scaling (DVS) is one of the most effective techniques to reduce energy consumption [20, 79, 82, 110]. In many microprocessor systems, the supply voltage can be changed by mode-set instructions according to the workload at run-time. With the trend of multiple cores being widely used in DSP systems, it is important to study DVS techniques for multiprocessor DSP systems. This chapter focuses on minimizing expected energy consumption with guaranteed probability satisfying timing constraints via DVS for real-time multiprocessor DSP systems.

In this chapter, we use probabilistic design space exploration and DVS to avoid over-designing systems. We propose two novel optimal algorithms, one for uniprocessor and one for multiprocessor DSP systems, to minimize the expected value of total energy consumption while satisfying timing constraints with guaranteed probabilities for real-time applications. Our work is related to the work in [37, 38]. In [37, 38], Hua et al. proposed a heuristic algorithm for uniprocessor and the *Probabilistic Data Flow Graph* (PDFG) is a simple path. We call the offline part of it as *HUA* algorithm for convenience. We also apply the greedy method of *HUA* algorithm to multiprocessor and call the new algorithm produced as *Heu*. We will compare our algorithms with Hua's algorithm for uniprocessor and *Heu* algorithm.

Our contributions are listed as the following:

- When there is a uniprocessor, the results of our algorithm, *VAP_S*, give the optimal solution and achieve significant energy saving than *HUA* algorithm.

- For the general problem, that is, when there are multiple processors and the PDFG is a DAG, our algorithm, *VAP_M*, gives the optimal solution and achieves significant average energy reduction than *Heu* algorithm.

- Our algorithms not only are optimal, but also provide more choices of smaller expected value of total energy consumption with guaranteed confidence probabilities satisfying timing constraints. In many situations, algorithms *HUA* and *Heu* cannot find a solution, yet ours can find satisfied results.

- Our algorithms are practical and quick. In practice, when the number of multi-parent nodes and multi-child nodes in the given PDFG graph is small, and the timing constraint is polynomial to the size of PDFG, the running times of these algorithms are very small and our experiments always finished in very short time.

We conduct experiments on a set of benchmarks, and compare our algorithms with *HUA* and *Heu* algorithms Experiments show that our algorithm for uniprocessor, *VAP_S*, has an average 58.0% energy-saving improvement with probability 0.8 satisfying timig constraint compared with the greedy algorithm *HUA*. Our algorithm for multiprocessor and DAG, *VAP_M*, has an average 56.1% energy-saving improvement compared with the results of the heuristic algorithm *Heu*.

The rest of the chapter is organized as following: The models and basic concepts are introduced in Section 3.2. In Section3.3, we give motivational examples. In Section 3.4, we propose our algorithms. The experimental results are shown in Section 3.5, and the conclusion is shown in Section 3.6.

3.2 Models and Concepts

In this section, we introduce the system model, the energy model, and VAP problem that will be used in the later sections.

System Model:

We focus on real-time applications on single-processor and multiprocessor DSP systems. *Probabilistic Data-Flow Graph* (PDFG) is used to model a embedded systems application. A *PDFG G* = $\langle U, ED, T, V \rangle$ is a *directed acyclic graph* (DAG), where $U = \langle u_1, \cdots, u_i, \cdots, u_N \rangle$ is a set of nodes; $V = \langle V_1, \cdots, V_j, \cdots, V_M \rangle$ is a voltage set; the execution time $T_{V_j}(u)$ is a random variable; $ED \subseteq U \times U$ is the edge set that defines the precedence relations among nodes in U. There is a timing constraint L and it must be satisfied for executing the whole PDFG. In the multiprocessor system, each processor has multiple discrete levels of voltages and its voltage level can be changed independently by voltage-level-setting instructions without the influence for other processors.

Energy Model:

Dynamic power, which is the dominant source of power dissipation in CMOS circuit, is proportional to $N \times C_s \times V_{dd}^2$, where N represent the number of computation cycles for a node, C_s is the effective switched capacitance, and V_{dd} is the supply voltage. Reducing the supply voltage can result in substantial power and energy saving. Roughly speaking, system's power dissipation is halved if we reduce V_{dd} by 30% without changing any other system parameters. However, this saving comes at the cost of reduced throughput, slower system clock frequency, or higher cycle period time (gate delay).

We use the similar energy model as in [79, 82, 110]. Let T represent the execution time of a node and E stand for energy consumption. The cycle period time T_c is proportional to $\frac{V_{dd}}{(V_{dd}-V_{th})^\alpha}$, where V_{th} is the threshold voltage and $\alpha \in (1.0, 2.0]$ is a technology dependent constant. Given the number of cycles N_c of node u, the supply voltage V_{dd} and the threshold voltage V_{th}, its computation time $T(u)$ and the energy $E(u)$ for node u are calculated as follows:

$$T_c = \frac{k \times V_{dd}}{(V_{dd} - V_{th})^\alpha} \tag{3.1}$$

$$T(u) = N_c \times T_c = N_c \times \frac{k \times V_{dd}}{(V_{dd} - V_{th})^\alpha} \tag{3.2}$$

$$E(u) = N_c \times C_s \times V_{dd}^2 \tag{3.3}$$

In Equation (3.1), k is a device related parameter. From Equations (3.2) and (3.3), we can see that the lower voltage will prolong the execution time of a node but reduce its energy consumption. In this paper, we assume that there is no energy or delay penalty associated with voltage switching and the energy leakage is very small.

DVS (Dynamic voltage scaling) is a technique that varies system's operating voltages and clock frequencies based on the computation load to provide desired performance with the minimum energy consumption. This technique has been demonstrated as one of the most effective low power system design techniques. Also, as we known, it has been supported by many modern microprocessors. There are many examples, such as Transmeta's Crusoe, AMD's K-6, Intel's XScale and Pentium III and IV, and some DSPs developed in Bell Labs [38].

Low power design is of particular interest for the soft real-time multimedia systems and we assume that our system has multiple voltages available on the chip such that the system can switch from one level to another. For the energy and delay overhead associated with multiple-voltage systems, we assume that voltage scaling occurs simultaneously without any energy and delay overhead.

VAP problem:

For multiple-voltage systems, assume there are maximum M different voltages in a voltage set $V = \langle V_1, V_2, \cdots, V_M \rangle$. For each voltage, there are maximum K execution time variations, although each node may have different execution time variations. An assignment for a PDFG G is to assign a processor to each node. Define an *assignment A* to be a function from domain U to range V, where U is the node set and V is the voltage set. For a node $u \in U$, $A(u)$ gives selected voltage level of node u.

In a PDFG G, each varied execution time is modeled as a probabilistic random variable, $T_{V_j}(u)$, $1 \leq j \leq M$, represents the execution times of each node $u \in U$ when running at voltage level V_j; $P_{V_j}(u)$, $1 \leq j \leq M$, represents the corresponding probability function. For each voltage V_j with respect to node u, there is a set of pairs (P_i, E_i), $\sum P_i = 1$, where P_i is the probability and E_i is the energy consumption of each node in PDFG. $E_{V_j}(u)$, $1 \leq j \leq M$, is used to represent the *expected value* of energy consumption of each node $u \in U$ at voltage V_j, $E_{V_j}(u) = \sum P_i E_i$, which is a fixed value. Because of the linearity of expected values, we can define the total energy consumption for a system. Given an assignment A of a PDFG G, we define the *system expected total energy consumption under assignment A*, denoted as $E_A(G)$, to be the summation of energy consumptions, $E_{A(u)}(u)$, $u \in U$, of all nodes, that is, $E_A(G) = \sum_{u \in U} E_{A(u)}(u)$. In this paper we call $E_A(G)$ *total energy consumption* in brief.

For the input PDFG G, given an assignment A, assume that $T_A(G)$ stands for the *execution time of graph G under assignment A*. $T_A(G)$ can be obtained from the longest path p in G. The new variable $T_A(G) = \max_{\forall p} T_{A(u)}(p)$, where $T_{A(u)}(p) = \sum_{v \in p} T_{A(u)}(u)$, is also a random variable. The *minimum expected total energy consumption E with con-*

fidence probability P under timing constraint L is defined as $E = \min_A E_A(G)$, where probability of ($T_A(G) \leq L$) $\geq P$. For each timing constraint L, our algorithm will output a serial of (Probability, Energy) pairs (P, E).

$T_{V_j}(u)$ is either a discrete random variable or a continuous random variable. We define $F(t)$ to be the *cumulative distribution function* (abbreviated as *CDF*) of the random variable $T_{V_j}(u)$, where $F(t) = P(T_{V_j}(u) \leq t)$. When $T_{V_j}(u)$ is a discrete random variable, the CDF $F(t)$ is the sum of all the probabilities associating with the execution times that are less than or equal to t. If $T_{V_j}(u)$ is a continuous random variable, then it has a *probability density function (PDF)* . If assume the PDF is f, then $F(t) = \int_0^t f(s)ds$. Function $F(t)$ is nondecreasing, and $F(-\infty) = 0$, $F(\infty) = 1$.

We define the *VAP (voltage assignment with probability) problem* as follows: Given M different voltage levels: $V_1, V_2, \cdots, V_M$, a PDFG $G = \langle U, ED \rangle$ with $T_{V_j}(u)$, $P_{V_j}(u)$, and $E_{V_j}(u)$ for each node $u \in U$ executed on each voltage V_j, a timing constraint L and a confidence probability P, find the voltage for each node in assignment A that gives the *minimum expected total energy consumption E with confidence probability P under timing constraint L*.

3.3   Motivational Example

3.3.1   Multiprocessor Systems

First we give an example for multiprocessor embedded systems, which is shown in Figure 3.1. In this paper, we assume the tasks have already been preprocessed. We already know which task will be processed by which processor, and the scheduling graph is given. For example, the task graph is shown in Figure 3.1(a). After preprocessing, we get the scheduling graph in Figure 3.1(c). This is a same graph to the input PDFG that is shown in Figure 3.2(b).

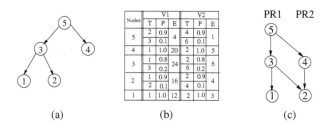

Figure 3.1. (a) A given PDFG. (b) The times, probabilities, and energy consumption of its nodes for different voltage levels. (c) The schedule graph of two processors.

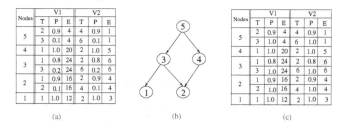

Figure 3.2. (a) The times, probabilities, and energy consumption of nodes for different voltage levels. (b) A DAG. (c) The time cumulative distribution functions (CDFs) and energy consumption of its nodes for different voltage levels.

In Figure 3.2, each node has two different voltage levels to choose from, and is executed on them with probabilistic execution times. The input PDFG has five nodes. The times, energy consumption, and probabilities of each node is shown in Figure 3.2(a). Node 5 is a multi-child node, which has two children: 3 and 4. Node 2 is a multi-parent node, and has two parents: 3 and 4. Figure 3.2(c) shows the time cumulative distribution functions (CDFs) and energy consumption of each node for different voltage levels. In DSP applications, a real-time system does not always has hard deadline time. The execution time can be smaller than the hard deadline time with certain probabilities. So, the hard deadline time is the worst-case of the varied smaller time cases. If we consider these time variations, we can achieve a better minimum energy consumption with satisfying confidence probabilities under timing constraints.

The number of computation cycles (N_c) for a task is proportional to the execution time. The energy consumption (E) depends on not only the voltage level V, but also the number of computation cycles N_c. We use the expected value of energy consumption ($Exp(E)$) as the energy consumption E under a certain voltage level V. Under different voltage levels, a task has different expected energy consumptions. The higher the voltage level is, the faster the execution time is, and the more expected energy is consumed. According to the energy model of DVS [41, 82, 110], the computation time is proportional to $V_{dd}/(V_{dd} - V_{th})^2$, where V_{dd} is the supply voltage, V_{th} is the threshold voltage; the energy consumption is proportional to V_{dd}^2. So here we assume the computation time of a node under the low voltage (V_2) is twice as much as it is under the high voltage (V_1); the energy consumption of a node under the high voltage (V_1) is four times as much as it is under the low voltage (V_2).

Table 3.1. Minimum expected total energy consumption with computed confidence probabilities under various timing constraints for a DAG.

T	(P , E)	(P , E)	(P , E)	(P , E)
4	0.65, 76			
5	0.65, 43	0.72, 61		
6	0.72, 34	0.81, 61		
7	0.80, 34	0.90, 52		
8	0.72, 22	0.80, 34	1.00, 52	
9	0.80, 22	0.90, 40	1.00, 52	
10	0.72, 19	0.80, 22	0.90, 34	1.00, 40
11	0.72, 19	0.80, 22	1.00, 34	
12	0.80, 19	0.90, 22	1.00, 34	
13	0.80, 19	1.00, 22		
14	0.90, 19	1.00, 22		
15	0.90, 19	1.00, 22		
16	1.00, 19			

3.3.2 Our Solution

For Figure 3.2, the minimum total energy consumptions with computed confidence proba-

bilities under the timing constraint are shown in Table 3.1. For each row of the table, the

E in each (P, E) pair gives the minimum total energy consumption with confidence prob-

ability P under timing constraint T. Using our algorithm, *VAP_M*, at timing constraint 11,

we can get (0.80, 22) pair. Table 3.2 shows the assignments of our algorithm. Assignment

$A(u)$ represents the voltage selection of each node u. Using our algorithm, we achieve

minimum total energy consumption 22 with probability 0.80 satisfying timing constraint

11. While using the heuristic algorithm *Heu*, the total energy consumption obtained is 61,

because *Heu* still need to use voltage level V_1 and cannot change all node's voltage level to

V_2 under timing constraint 11. The energy saving improvement of our algorithm is 59.1%.

This case shows that in many situations, the solutions obtained by our algorithms have

significant improvement compared with the results gotten by *Heu*.

Table 3.2. Under timing constraint 11, the different assignments between *VAP_M* and *Heu*.

		Node id	T	V Level	Prob.	Energy
Ours	$A(u)$	5	3	V1	1.00	4
		4	2	V2	1.00	5
		3	4	V2	0.80	6
		2	2	V2	1.00	4
		1	4	V2	1.00	3
	Total		11		0.80	22
HUA	$A(u)$	5	3	V1	1.00	4
		4	2	V1	1.00	5
		3	1	V1	0.80	24
		2	1	V1	1.00	12
		1	2	V1	1.00	16
	Total		6		0.80	61

Given the requirement of probability, we can get a minimum total energy consump-

tion under every timing constraint. For example, if the required probability is 0.80, then the total energy consumption varies from 61 to 19 at different timing constraints. The energy consumptions under different timing constraints are shown in Table 3.3.

Table 3.3. Given an requiremnt of probability, the (Probability, Energy) pairs under different timing constraints.

T	6	7 , 8	9 - 11	12 - 16
(P , E)	(0.80, 61)	(0.80, 34)	(0.80, 22)	(0.80, 19)

3.4   The Algorithms

3.4.1   Definitions and Lemma

To solve the VAP problem, we use dynamic programming method traveling the graph in bottom up fashion. For the ease of explanation, we will index the nodes based on bottom up sequence. For example, Figure 3.1 (a) shows a tree indexed by bottom up sequence. The sequence is: $u_1 \to u_2 \to \cdots \to u_5$. Define a *root* node to be a node without any parent and a *leaf* node to be a node without any child. A multi-child node is a node with more than one child. For example, in Figure 3.1 (a), node 3 and 5 are multi-child nodes. Similarly, a multi-parent node is a node with more than one parent.

Given the timing constraint L, a PDFG G, and an assignment A, we first give several definitions as follows:

- $\mathbf{G^i}$: The sub-graph rooted at node u_i, containing all the nodes reached by node u_i. In our algorithm, each step will add one node which becomes the root of its sub-graph. For example, in Figure 3.1 (a), G^i is the tree containing nodes 1, 2, and 3.

- $\mathbf{E_A(G^i)}$ and $\mathbf{T_A(G^i)}$: The total energy consumption (E) and total execution time (T) of G^i under the assignment A. In our algorithm, each step will achieve the minimum total E of G^i with computed confidence probabilities under various timing constraints.

- In our algorithm, table $D_{i,j}$ will be built. Each entry of table $D_{i,j}$ will store a linked list of (Probability, Energy) pairs sorted by probability in ascending order. Here we define the **(Probability, Energy) pair ($P_{i,j}$, $E_{i,j}$)** as follows: $E_{i,j}$ is the minimum energy consumption of $E_A(G^i)$ computed by all assignments A satisfying $T_A(G^i) \leq j$ with probability $\geq P_{i,j}$.

We introduce the *operator* "$\oplus$" in this paper. For two (Probability, Energy) pairs H_1 and H_2, if H_1 is $(P^1_{i,j}, E^1_{i,j})$, and H_2 is $(P^2_{i,j}, E^2_{i,j})$, then, after the $\oplus$ operation between H_1 and H_2, we get pair (P', C'), where $P' = P^1_{i,j} * P^2_{i,j}$ and $C' = E^1_{i,j} + E^2_{i,j}$. We denote this operation as "$H_1 \oplus H_2$". This is the key operation of our algorithms. The meaning is that when two task nodes add together, the total energy is computed by adding the energy of all nodes together and the probability corresponding to the total energy is computed by multiplying the probabilities of all nodes based on the basic properties of probability and energy of a PDFG. As we know, for two independent events A and B, $P(A \cup B) = P(A) * P(B)$, and $C(A \cup B) = C(A) + C(B)$.

In our algorithm, $D_{i,j}$ is the table in which each entry has a linked list that store pair $(P_{i,j}, E_{i,j})$. Here, i represents a node number, and j represents time. For example, a linked list can be $(0.1, 2) \rightarrow (0.3, 3) \rightarrow (0.8, 6) \rightarrow (1.0, 12)$. There is no redundant pairs in this linked list. But usually, there are redundant pairs in a linked list. The redundant-pair removal algorithm is shown in Algorithm 3.4.1.

For example, we have a list with pairs $(0.1, 2) \rightarrow (0.3, 3) \rightarrow (0.5, 3) \rightarrow (0.3, 4)$, we do the redundant-pair removal as following: First, sort the list according $P_{i,j}$ in an ascending order. This list becomes to $(0.1, 2) \rightarrow (0.3, 3) \rightarrow (0.3, 4) \rightarrow (0.5, 3)$. Second,

---

Algorithm 3.4.1 Redundant-pair removal algorithm

Input: A list of $(P_{i,j}^k, E_{i,j}^k)$
Output: A redundant-pair free list

1: Sort the list by $P_{i,j}$ in an ascending order such that $P_{i,j}^k \leq P_{i,j}^{k+1}$.
2: From the beginning to the end of the list,
3: **for** each two neighboring pairs $(P_{i,j}^k, E_{i,j}^k)$ and $(P_{i,j}^{k+1}, E_{i,j}^{k+1})$ **do**
4:    **if** $P_{i,j}^k = P_{i,j}^{k+1}$ **then**
5:       **if** $E_{i,j}^k \geq E_{i,j}^{k+1}$ **then**
6:          cancel the pair $P_{i,j}^k, E_{i,j}^k$
7:       **else**
8:          cancel the pair $P_{i,j}^{k+1}, E_{i,j}^{k+1}$
9:       **end if**
10:   **else**
11:       **if** $E_{i,j}^k \geq E_{i,j}^{k+1}$ **then**
12:          cancel the pair $(P_{i,j}^k, E_{i,j}^k)$
13:       **end if**
14:   **end if**
15: **end for**

---

cancel redundant pairs. Comparing $(0.1, 2)$ and $(0.3, 3)$, we keep both. For the two pairs $(0.3, 3)$ and $(0.3, 4)$, we cancel pair $(0.3, 4)$ since the cost 4 is bigger than 3 in pair $(0.3, 3)$. Comparing $(0,3, 3)$ and $(0.5, 3)$, we cancel $(0.3, 3)$ since $0.3 < 0.5$ while $3 \geq 3$. There is no information lost in redundant-pair removal.

In summary, we can use the following Lemma to cancel redundant pairs, which is shown in Lemma 3.4.1.

Lemma 3.4.1. *Given $(P_{i,j}^1, E_{i,j}^1)$ and $(P_{i,j}^2, E_{i,j}^2)$ in the same list:*

1. *If $P_{i,j}^1 = P_{i,j}^2$, then the pair with minimum $E_{i,j}$ is selected to be kept.*

2. *If $P_{i,j}^1 < P_{i,j}^2$ and $E_{i,j}^1 \geq E_{i,j}^2$, then $E_{i,j}^2$ is selected to be kept.*

Using Lemma 3.4.1, we can cancel many redundant-pair $(P_{i,j}, E_{i,j})$ whenever we find conflicting pairs in a list during a computation. After the $\oplus$ operation and redundant pair removal, the list of $(P_{i,j}, E_{i,j})$ has the following properties:

Lemma 3.4.2.   *1. $P_{i,j}^1 \neq P_{i,j}^2$ and $E_{i,j}^1 \neq E_{i,j}^2$.*

2. $P_{i,j}^1 < P_{i,j}^2$ if and only if $E_{i,j}^1 < E_{i,j}^2$.

Since the linked list is in an ascending order by probabilities, if $P_{i,j}^1 < P_{i,j}^2$, while $E_{i,j}^1 \geq E_{i,j}^2$, based on the definitions, we can guarantee with $P_{i,j}^1$ to find smaller energy consumption $E_{i,j}^2$. Hence $(P_{i,j}^2, E_{i,j}^2)$ has already covered $(P_{i,j}^1, E_{i,j}^1)$. We can cancel the pair $(P_{i,j}^1, E_{i,j}^1)$. For example, we have two pairs: (0.1, 3) and (0.6, 2). Since (0.6, 2) is better than (0.1, 3), in other words, (0.6, 2) covers (0.1, 3), we can cancel (0.1, 3) and will not lose useful information. We can prove the vice versa is true similarly. When $P_{i,j}^1 = P_{i,j}^2$, smaller E is selected.

In every step in our algorithm, one more node will be included for consideration. The information of this node is stored in local table $B_{i,j}$, which is similar to table $D_{i,j}$. A local table store only data of probabilities and energy of a node itself. Table $B_{i,j}$ is the local table storing only the data of node u_i. In more detail, $E_{i,j}$ is a local table of linked lists that store pair $(P_{i,j}, E_{i,j})$ sorted by $P_{i,j}$ in an ascending order; $E_{i,j}$ is the cost only for node u_i at time j, and $P_{i,j}$ is the corresponding probability. The building procedures of $B_{i,j}$ are as follows. First, sort the execution time variations in an ascending order. Then, accumulate the probabilities of same type. Finally, let $L_{i,j}$ be the linked list in each entry of $E_{i,j}$, insert $L_{i,j}$ into $L_{i,j+1}$ while redundant pairs canceled out based on Lemma 3.4.1. For example, if a node has the following (T: P, E) pairs: (1: 0.9, 10), (3: 0.1, 10) for type V_1, and (2: 0.7, 4), (4: 0.3, 4) for type V_2. After sorting and accumulating, we get (1: 0.9, 10), (2: 0.7, 4), (3: 1.0, 10), and (4: 1.0, 4). We obtain Table 3.4 after the insertion.

Similarly, for two linked lists L_1 and L_2, the operation "$\mathbf{L_1} \oplus \mathbf{L_2}$" is implemented as follows: First, implement $\oplus$ operation on all possible combinations of two pairs from different linked lists. Then insert the new pairs into a new linked list and remove redundant-pair using Lemma 3.4.1.

Table 3.4. An example of local table, $B_{0,j}$

$Time$	1	2	3	4
(P_i, E_i)	(0.9, 10)	(0.7, 4)	(0.7, 4)	(1.0, 4)
		(0.9, 10)	(1.0, 10)	

For an input PDFG with multiple processors, given the order of nodes and expected energy consumption of each node, the basic steps of our schedule are shown in algorithm *VAP_SG*.

Algorithm 3.4.2 Algorithm to get scheduling graph (*VAP_SG*)

Input: a task graph PDFG
Output: a scheduling graph
  1: Build a graph to show the order using list scheduling.
  2: Show the dependency in the graph.
  3: Remove all redundant edges according Lemma 3.4.3.

In the graph built in step 1 of *VAP_SG*, If there is an edge from u_i to u_j, this means that u_i is scheduled before u_j in the same processor or u_j depends on u_i in the original PDFG. The new graph is a DAG that represents the order of nodes and dependencies.

Lemma 3.4.3. *For two nodes u_i and u_j, there is an edge e_{ij}, if we can find another separate path $u_i \rightarrow u_j$, then the edge e_{ij} can be deleted.*

For example, Figure 3.3 shows the input PDFG. There are two processors PR_1 and PR_2. The order of nodes are given. In Figure 3.3(b), there are paths $C \rightarrow E$ amd $E \rightarrow G$. Also, there is a separate path $C \rightarrow G$. Then we can cancel the path $C \rightarrow G$.

3.4.2 The VAP_S Algorithm

The algorithm for uniprocessor system is shown in *VAP_S*. It can give the optimal solution for the VAP problem when there is only one processor.

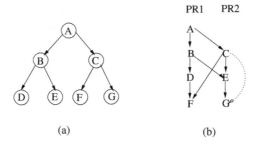

(a)   (b)

Figure 3.3. (a) A task graph with seven nodes. (b) The schedule graph for two processors using list scheduling.

The VAP_S Algorithm

In algorithm *VAP_S*, we first build a local table $B_{i,j}$ for each node. Next, in step 2 of the algorithm, when $i = 1$, there is only one node. We set the initial value, and let $D_{1,j} = B_{1,j}$. Then using dynamic programming method, we build the table $D_{i,j}$. For each node u_i under each time j, we try all the times k ($1 \leq k \leq j$) in table $B_{i,j}$. We use "$\oplus$" on the two tables $B_{i,k}$ and $D_{i-1,j-k}$. Since $k + (j - k) = j$, the total time of nodes from u_1 to u_i is j. The "$\oplus$" operation adds the energy consumptions of two tables together and multiplies the probabilities of two tables with each other. Finally, we use Lemma 3.4.1 to cancel the conflicting (Probability, Energy) pairs. The new energy consumption in each pair obtained in table $D_{i,j}$ is the energy consumption of current node u_i at time k plus the energy consumption in each pair obtained in $D_{i-1,j-k}$. Since we have already used Lemma 3.4.1 canceling redundant pairs, the energy consumption of each pair in $D_{i,j}$ is the minimum total energy consumption for graph G^i with confidence probability $P_{i,j}$ under timing constraint j.

The energy consumption in $D_{N,j}$ is the minimum total energy consumption with computed confidence probability under timing constraint j. Given the timing constraint L,

---

Algorithm 3.4.3 Optimal algorithm for the VAP problem when there is a single processor (*VAP_S*)

Input: M different levels of voltages, a DAG, and the timing constraint L.
Output: An optimal voltage level assignment

1. Build a local table $B_{i,j}$ for each node of PDFG.

2.   1: let $D_{1,j} = B_{1,j}$
     2: **for** each node $u_i, i > 1$ **do**
     3:    **for** each time j **do**
     4:      **for** each time k in $B_{i,k}$ **do**
     5:        **if** $D_{i-1,j-k}! = NULL$ **then**
     6:          $D_{i,j} = D_{i-1,j-k} \oplus B_{i,k}$
     7:        **else**
     8:          continue
     9:        **end if**
     10:     **end for**
     11:     insert $D_{i,j-1}$ to $D_{i,j}$ and remove redundant pairs using Lemma 3.4.1 and redundant-pair remove algorithm.
     12:   **end for**
     13: **end for**

3. return $D_{N,j}$

---

the minimum total energy consumption for the graph G is the energy consumption in $D_{N,L}$. In the following, we show Theorem 3.4.1 about this.

Theorem 3.4.1. *For each pair $(P_{i,j}, E_{i,j})$ in $D_{i,j}$ $(1 \leq i \leq N)$ obtained by algorithm VAP_S, $E_{i,j}$ is the minimum total energy consumption for graph G^i with confidence probability $P_{i,j}$ under timing constraint j.*

Proof. By induction. **Basic Step:** When $i = 1$, there is only one node and $D_{1,j} = B_{1,j}$. $D_{1,j} = \min_{1 \leq k \leq M} \{c_k(1) \text{ if } j \geq t_k(1)\}$. Thus, Thus, when $i = 1$, Theorem 3.4.1 is true.
Induction Step: We need to show that for $i \geq 1$, if for each pair $(P_{i,j}, E_{i,j})$ in $D_{i,j}$, $E_{i,j}$ is the minimum total energy consumption for graph G^i with confidence probability $P_{i,j}$ under timing constraint j, then for each pair $(P_{i+1,j}, E_{i+1,j})$ in $D_{i+1,j}$, $E_{i+1,j}$ is the minimum total energy consumption for graph G^{i+1} with confidence probability $P_{i+1,j}$ under timing

constraint j. In step 2 of the algorithm, since $j = k + (j - k)$ for each k in $B_{i+1,j}$, we try all the possibilities to obtain j. Then we use $\oplus$ operator to add the energy consumptions of two tables and multiply the probabilities of two tables. Finally, we use Lemma 3.4.1 to cancel the conflicting (Probability, Energy) pairs. The new energy consumption in each pair obtained in table $D_{i+1,j}$ is the energy consumption of current node $i + 1$ at time k plus the energy consumption in each pair obtained in $D_{i,j-k}$. Since we have used Lemma 3.4.1 to cancel redundant pairs, the energy consumption of each pair in $D_{i+1,j}$ is the minimum total energy consumption for graph G^{i+1} with confidence probability $P_{i+1,j}$ under timing constraint j. Thus, Theorem 3.4.1 is true for any i ($1 \leq i \leq N$). $\qquad\square$

From Theorem 3.4.1, we know $D_{N,L}$ records the minimum total energy consumption of the whole path with corresponding confidence probabilities under the timing constraint L. We can record the corresponding FU type assignment of each node when computing the minimum total energy consumption in step 2 in the algorithm VAP_S. Using these information, we can get an optimal assignment by tracing how to reach $D_{N,L}$.

It takes $O(M * K)$ to compute one value of $D_{i,j}$, where M is the maximum number of FU types, and K is the maximum number of execution time variations for each node. Thus, the complexity of the algorithm VAP_S is $O(|U| * L * M * K)$, where $|U|$ is the number of nodes and L is the given timing constraint. Usually, the execution time of each node is upper bounded by a constant. So L equals $O(|U|^c)$ (c is a constant). In this case, VAP_S is polynomial.

3.4.3 The VAP_M Algorithm

In this subsection, we give a heuristic algorithm Heu first, then we propose our novel and optimal algorithm, VAP_M, for multiprocessor DSP systems. We will compare them in the experiments section.

The Heu Algorithm

We first design an heuristic algorithm for multiprocessor systems according to the *HUA* algorithm in [37, 38], we call this algorithm as *Heu*. The PDFG now is a DAG and no longer limited to a simple path. The authors of [37, 38] did not give the algorithm for multiple processors situation, and their data flow graph is a simple path. *Heu* is an algorithm that use the idea of the algorithm of [37, 38] and using for multiple processors situation, the data flow graph is a DAG.

Algorithm 3.4.4 Heuristic algorithm for the VAP problem when there are multiple processors and the DFG is DAG (*Heu*)

Input: M different levels of voltages, a DAG, and the timing constraint L.

Output: a voltage assignment to minimize energy consumption with a guaranteed probability θ satisfying L

1: Schedue graph construction.
2: /* assign worst-case time to u_i */
   for each vertex u_i, let $l_i = k_i$;
3: $P = 1$;
4: while $(P > \theta)$
5: {
6:    pick u_i that has the maximum $(t_{il_i} - t_{i(l_i-1)}) \cdot \frac{P_{il_i}}{P_{i(l_i-1)}}$;
7:    $P = P \cdot \frac{P_{il_i}}{P_{i(l_i-1)}}$;
8:    if $(P > \theta)$
9:        $l_i = l_i - 1$;
10: }
11: /* calculate the total execution time T */
    $T = \sum t_{il_i}$;
12: /* if θ cannot be met */
    if $(T > L)$ exit;
13: for each vertex u_i, let $S_i = t_{il_i} \cdot \theta/T$;

In *Heu* algorithm, k_i is the largest possible time variation for node u_i, l_i is the time variation of node u_i, S_i is the scaled time slot for node u_i, and t_{ij} is the j time variation of node u_i. Let $E_{ij}(S_i)$ be the minimum energy to complete the workload t_{ij} required by vertex u_i in time S_i. On an ideal variable voltage processor, $E_{ij}(S_i)$ is the energy consumed by running at voltage V_{ideal} throughout the entire assigned slot S_i, where V_{ideal} is the voltage level that enables the processor to accumulate the workload t_{ij} at the end of the slot. On a multiple voltage processor with only a finite set of voltage levels $V_1 < V_2 < \cdots$, $E_{ij}(S_i)$ is the energy consumed by running at V_k for a certain amount of time and then switching to the next higher level V_{k+1} to complete t_{ij}, where $V_k < V_{ideal} < V_{k+1}$ and the switching point can be conveniently calculated from S_i and t_{ij} [37, 38].

We propose our algorithm, *VAP_M*, for multiprocessor DSP systems, which shown as follows. In *VAP_M*, we exhaust all the possible assignments of multi-parent or multi-child nodes. Without loss of generality, assume we using bottom up approach. If the total number of nodes with multi-parent is t, and there are maximum K variations for the execution times of all nodes, then we will give each of these t nodes a fixed assignment.

The VAP_M Algorithm

Input: M different voltage levels, a DAG, and the timing constraint L.

Output: An optimal voltage assignment for the DAG

1. Topological sort all the nodes, and get a sequence A.

2. Count the number of multi-parent nodes t_{mp} and the number of multi-child nodes t_{mc}. If $t_{mp} < t_{mc}$, use bottom up approach; Otherwise, use top down approach.

3. For bottom up approach, use the following algorithm. For top down approach, just reverse the sequence.

4. If the total number of nodes with multi-parent is t, and there are maximum K variations for the execution times of all nodes, then we will give each of these t nodes a fixed assignment.

5. For each of the K^t possible fixed assignments, Assume the sequence after topological sorting is $u_1 \rightarrow u_2 \rightarrow \cdots \rightarrow u_N$, in bottom up fashion. Let $D_{1,j} = B_{1,j}$.

Assume $D'_{i,j}$ is the table that stored minimum total energy consumption with computed confidence probabilities under the timing constraint j for the sub-graph rooted on u_i except u_i. Nodes $u_{i_1}, u_{i_2}, \cdots, u_{i_w}$ are all child nodes of node u_i and w is the number of child nodes of node u_i, then

$$D'_{i,j} = \begin{cases} (0,0) & \text{if } w = 0 \\ D_{i_1,j} & \text{if } w = 1 \\ D_{i_1,j} \oplus \cdots \oplus D_{i_w,j} & \text{if } w > 1 \end{cases} \qquad (3.4)$$

6. For $D_{i_1,j} \oplus D_{i_2,j}$, G' is the union of all nodes in the graphs rooted at nodes u_{i_1} and u_{i_2}. Travel all the graphs rooted at nodes u_{i_1} and u_{i_2}. If a node is a common node, then use a selection function to choose the type of a node.

7. Then, for each k in $B_{i,k}$.

$$D_{i,j} = D'_{i,j-k} \oplus B_{i,k} \qquad (3.5)$$

8. For each possible fixed assignment, we get a $D_{N,j}$. Merge the (Probability, Energy) pairs in all the possible $D_{N,j}$ together, and sort them in ascending sequence according probability.

9. Then use the Lemma 3.4.1 to remove redundant pairs. Finally get $D_{N,j}$.

Now we explain our optimal algorithm *VAP_M* in details. In *VAP_M*, we exhaust all the possible assignments of multi-parent or multi-child nodes. Without loss of generality, assume we using bottom up approach. If the total number of nodes with multi-parent is t, and there are maximum K variations for the execution times of all nodes, then we will give each of these t nodes a fixed assignment. We will exhaust all of the K^t possible fixed assignments.

Algorithm *VAP_M* gives the optimal solution when the given PDFG is a DAG. In the following, we give the Theorem 3.4.2 and Theorem 3.4.3 about this.

Theorem 3.4.2. *In each possible fixed assignment, for each pair $(P_{i,j}, E_{i,j})$ in $D_{i,j}$ $(1 \leq i \leq N)$ obtained by algorithm VAP_M, $E_{i,j}$ is the minimum total energy consumption for the graph G^i with confidence probability $P_{i,j}$ under timing constraint j.*

Proof. By induction. **Basic Step:** When $i = 1$, There is only one node and $D_{1,j} = B_{1,j}$. Thus, when $i = 1$, Theorem 3.4.2 is true. **Induction Step:** We need to show that for $i \geq 1$, if for each pair $(P_{i,j}, E_{i,j})$ in $D_{i,j}$, $E_{i,j}$ is the minimum total energy consumption of the graph G^i, then for each pair $(P_{i+1,j}, E_{i+1,j})$ in $D_{i+1,j}$, $E_{i+1,j}$ is the total energy consumption of the graph G^{i+1} with confidence probability $P_{i+1,j}$ while satisfying timing constraint j.

According to the bottom up approach (for top down approach, just reverse the sequence), the execution of $D_{i,j}$ for each child node of u_{i+1} has been finished before executing $D_{i+1,j}$. From equation (3.4), $D'_{i+1,j}$ gets the summation of the minimum total energy consumption of all child nodes of u_{i+1} because they can be executed simultaneously within time j. We avoid the repeat counting of the common nodes. Hence, each node in the graph rooted by node u_{i+1} was counted only once. From equation (3.5), the minimum total energy consumption is selected from all possible energy consumptions caused by adding u_{i+1} into the sub-graph rooted on u_{i+1}. So for each pair $(P_{i+1,j}, E_{i+1,j})$ in $D_{i+1,j}$, $E_{i+1,j}$ is the total energy consumption of the graph G^{i+1} with confidence probability $P_{i+1,j}$ under timing constraint j. Therefore, Theorem 3.4.2 is true for any i $(1 \leq i \leq N)$. $\square$

Theorem 3.4.3. *For each pair $(P_{i,j}, E_{i,j})$ in $D_{N,j}$ $(1 \leq j \leq L)$ obtained by algorithm VAP_M, $E_{i,j}$ is the minimum total energy consumption for the given DAG G with confidence probability $P_{i,j}$ under timing constraint j.*

Proof. According to Theorem 3.4.2, in each possible fixed assignment, for each pair $(P_{i,j}, E_{i,j})$ in $D_{i,j}$ we obtained, $E_{i+1,j}$ is the total energy consumption of the graph G^{i+1} with confidence probability $P_{i+1,j}$ under timing constraint j. In algorithm *VAP_M*, we try all the possible fixed assignments, combine them together into a new row $D_{N,j}$ in dynamic table, and remove redundant pairs using the Lemma 3.4.1. Hence, for each pair $(P_{i,j}, E_{i,j})$ in $D_{N,j}$ $(1 \leq j \leq L)$ obtained by algorithm *VAP_M*, $E_{i,j}$ is the minimum total energy consumption for the given DAG G with confidence probability $P_{i,j}$ while satisfying timing constraint j. □

In algorithm *VAP_M*, there are K^t loops and each loop needs $O(|U|^2 * L * M * K)$ running time. The complexity of algorithm *VAP_M* is $O(K^{t+1} * |U|^2 * L * M)$, where t is the total number of nodes with multi-parent (or multi-child) in bottom up approach (or top down approach), $|U|$ is the number of nodes, L is the given timing constraint, M is the maximum number of FU types for each node, and K is the maximum number of execution time variation for each node. Algorithm *VAP_M* is exponential, hence it can not be applied to a graph with large amounts of multi-parent and multi-child nodes.

3.5   Experiments

This section presents the experimental results of our algorithms. We conduct experiments on a set of benchmarks including 4-stage lattice filter, 8-stage lattice filter, voltera filter, differential equation solver, RLS-languerre lattice filter, and elliptic filter. Among them, the PDFG for first three filters are trees and those for the others are DAGs.

Three different voltage levels, V_1, V_2, and V_3, are used in the system, in which a processor under V_1 is the quickest with the highest energy consumption and a processor under V_3 is the slowest with the lowest energy consumption. The distribution of execution times of each node is Gaussian. For each benchmark, the first timing constraint we use is the minimum execution time. The experiments are performed on a Dell PC with a P4 2.1 G processor and 512 MB memory running Red Hat Linux 7.3.

Table 3.5. The minimum expected total energy consumption with computed confidence probabilities under various timing constraints for voltera filter.

Voltera Filter (27 nodes)									
TC	0.8			0.9			1.0		
	HUA	Ours	%	HUA	Ours	%	HUA	Ours	%
103	708	701	1.0	×	×		×	×	
108	708	647	8.6	708	701	1.0	×	×	
110	708	623	12.1	708	671	5.2	708	701	1.0
150	708	310	56.2	708	338	52.3	708	347	51.0
200	708	186	73.7	708	204	71.2	708	206	70.9
206	177	175	1.2	708	196	72.3	708	198	72.0
216	177	162	8.5	177	175	1.2	708	182	74.3
220	177	156	11.9	177	172	2.8	177	175	1.2
250	177	118	33.4	177	130	26.6	177	132	25.6
300	177	73	59.8	177	76	57.1	177	82	54.7
350	177	45	74.6	177	45	74.6	177	45	74.6
Ave. Redu.(%)		58.6			61.7			64.8	

We compare our uniprocessor algorithm with the heuristic algorithm *HUA* in [37, 38] on all of the six benchmarks. There is only one processor: $PR1$. These experiments are finished in less than one second. The experimental results on voltera filter, 4-stage lattice filter, and 8-stage lattice filter are shown in Table 3.5-3.7. In each table, column "TC" represents the given timing constraint, "HUA" represents the heuristic algorithm *HUA* in [37, 38], and "Ours" represents our optimal algorithm *VAP_S*. The minimum total energy consumption obtained from different algorithms: *VAP_S* and *HUA* [37, 38], are presented in each entry. Columns "1.0", "0.9", and "0.8", represent that the confidence probability is 1.0, 0.9, and 0.8, respectively.

Column "%" shows the percentage of reduction on the total energy consumption, compared the results of algorithm *HUA* [37,38]. The average percentage reduction is shown in the last row "Ave. Redu(%)" of all Tables 3.5-3.7, which is computed by averaging energy-savings at all different timing constraints. The entry with "×" means no solution available. For the timing constraint below certain value, we can not find a solution using *HUA* algorithm, while we can find solutions using our algorithm. In Table 3.5, for exam-

Table 3.6. The minimum expected total energy consumption with computed confidence probabilities under various timing constraints for 4-stage lattice filter.

4-stage Lattice IIR Filter (26 nodes)								
TC	0.8			0.9			1.0		
	HUA	Ours	%	HUA	Ours	%	HUA	Ours	%
96	684	676	1.0	×	×		×	×	
101	684	616	9.8	684	676	1.0	×	×	
107	684	592	13.5	684	671	6.7	684	676	1.0
150	684	309	54.8	684	355	48.2	684	369	45.8
180	684	212	69.0	684	227	66.8	684	238	65.2
192	171	168	1.8	684	186	62.9	684	190	62.2
202	171	154	10.0	171	168	1.8	684	174	64.6
214	171	150	12.1	171	152	11.8	171	168	1.8
240	171	108	26.9	171	120	29.9	171	124	27.5
280	171	64	62.6	171	68	60.2	171	74	56.7
320	171	41	76.0	171	41	76.0	171	41	76.0
Ave. Redu.(%)		55.9			59.7			62.4	

ple, under timing constraint 103, we can not find solution for probability 0.9 using *HUA* algorithm, but we can find solution 705 using our optimal algorithm.

From the Table 3.5-3.7, we found in many situations, algorithm *VAP_S* has significant energy consumption reduction than algorithm *HUA* [37,38]. For example, in Table 3.5, under the timing constraint 200, for probability 0.8, the entry under "HUA" is 708, which is the minimum total energy consumption using algorithm *HUA*. The entry under "Ours" is 186, which means by using *VAP_S* algorithm, we can achieve minimum total energy consumption 186 with confidence probability 0.8 under timing constraint 200. The energy reduction is 73.7%.

The experimental results show that our algorithm can greatly reduce the total energy consumption while have a guaranteed confidence probability satisfying timing constraints. On average, algorithm *VAP_S* gives an energy consumption reduction of 58.0% with 0.8 confidence probability satisfying timing constraints, and energy consumption reductions of 61.4% and 64.1% with 0.9 and 1.0 confidence probabilities satisfying timing constraints,

Table 3.7. The minimum expected total energy consumption with computed confidence probabilities under various timing constraints for 8-stage lattice filter.

| 8-stage Lattice IIR Filter (42 nodes) | | | | | | | | |
| TC | 0.8 | | | 0.9 | | | 1.0 | | |
	HUA	Ours	%	HUA	Ours	%	HUA	Ours	%
166	1100	1090	0.9	×	×		×	×	
172	1100	1003	8.2	1100	1090	0.9	×	×	
175	1100	960	12.8	1100	1032	6.2	1100	1090	0.9
200	1100	548	50.2	1100	597	45.7	1100	626	43.1
180	1100	404	63.2	1100	421	61.7	1100	442	59.8
332	275	272	1.1	1100	305	72.3	1100	308	72.0
344	275	250	9.0	275	272	1.1	1100	282	74.4
350	275	239	13.2	275	242	12.1	275	272	1.1
400	275	183	33.5	275	200	27.3	275	205	25.5
420	275	112	59.3	275	118	57.1	275	127	53.9
450	275	68	75.3	275	68	75.3	275	68	75.3
Ave. Redu.(%)		59.5			62.8			65.2	

respectively. The experiments using *VAP_S* algorithm on these benchmarks are finished within several minutes.

For the multiprocessor systems, we compare our *VAP_M* algorithm with the *Heu* algorithm. We also conduct experiments on all of the six benchmarks. There are two processors: $PR1$ and $PR2$. The experimental results for different equation solver, RLS-Laguerre filter, and elliptic filter are shown in Table 3.8-3.10. In each table, column "TC" represents the given timing constraint, "Heu" represents the heuristic algorithm *Heu*, and "Ours" represents our optimal algorithm *VAP_M*. The minimum total energy consumption obtained from different algorithms: *VAP_M* and *Heu*, are presented in each entry. Columns "1.0", "0.9", and "0.8", represent that the confidence probability is 1.0, 0.9, and 0.8, respectively.

Column "%" shows the percentage of reduction on the total energy consumption, compared the results of *Heu*. The average percentage reduction is shown in the last row "Ave. Redu(%)" of all Tables 3.8-3.10. The entry with "×" means no solution available. Under timing constraint 63 in Table 3.8, there is no solution for probability 0.9 using *Heu*.

Table 3.8. The minimum expected total energy consumption with computed confidence probabilities under various timing constraints for different equation solver.

Different Equation Solver (11 nodes)									
TC	0.8			0.9			1.0		
	Heu	Ours	%	Heu	Ours	%	Heu	Ours	%
63	432	426	1.5	×	×		×	×	
66	432	394	8.8	432	426	1.5	×	×	
68	432	380	12.1	432	397	8.2	432	426	1.5
60	432	196	54.6	432	208	51.8	432	214	50.4
80	432	162	72.6	432	126	70.8	432	133	69.2
126	108	104	3.7	432	118	62.7	432	123	71.6
132	108	98	9.2	108	104	3.7	432	112	74.1
138	108	95	12.5	108	103	4.2	108	104	3.7
150	108	72	33.4	108	80	26.0	108	84	22.3
180	108	46	57.4	108	48	55.6	108	58	46.3
200	108	32	70.4	108	32	70.4	108	32	70.4
Ave. Redu.(%)		52.6			55.3			57.4	

However, we can find solution 426 with 0.9 probability that guarantees the total execution time of the PDFG are less than or equal to the timing constraint 63.

From the Table 3.8-3.10, we found in many situations, algorithm *VAP_M* has significant energy consumption reduction than algorithm *Heu*. For example, in Table 3.8, under the timing constraint 80, for probability 0.8, the entry under "Heu" is 432, which is the minimum total energy consumption using *Heu*. The entry under "Ours" is 162, which means by using *VAP_M*, we can achieve minimum total energy consumption 186 with confidence probability 0.8 under timing constraint 80. The energy reduction is 72.6%.

The experimental results show that our algorithm can greatly reduce the total energy consumption while have a guaranteed confidence probability. On average, algorithm *VAP_M* gives an energy consumption reduction of 56.1% with 0.8 confidence probability satisfying timing constraints, and energy consumption reductions of 59.3% and 61.7% with 0.9 and 1.0 confidence probabilities satisfying timing constraints, respectively. The experiments using *VAP_M* on these benchmarks are finished within several minutes.

Table 3.9. The minimum expected total energy consumption with computed confidence probabilities under various timing constraints for RLS-Laguerre filter.

RLS-Laguerre Filter (19 nodes)									
TC	0.8			0.9			1.0		
	Heu	Ours	%	Heu	Ours	%	Heu	Ours	%
130	900	892	0.9	×	×		×	×	
136	900	804	10.6	900	892	0.9	×	×	
140	900	756	16.0	900	727	9.2	900	892	0.9
180	900	408	54.6	900	430	52.2	900	444	50.7
220	900	268	70.2	900	277	69.2	900	261	70.0
260	225	218	3.2	900	248	72.4	900	251	72.1
272	225	202	10.2	225	218	3.2	900	232	74.2
280	225	186	17.2	225	212	6.0	225	218	3.2
320	225	152	32.4	225	164	27.2	225	172	23.6
360	225	94	58.2	225	98	56.4	225	106	52.9
420	225	58	74.2	225	58	74.2	225	58	74.2
Ave. Redu.(%)		54.3				57.6			59.2

Table 3.10. The minimum expected total energy consumption with computed confidence probabilities under various timing constraints for elliptic filter.

Elliptic Filter (34 nodes)									
TC	0.8			0.9			1.0		
	Heu	Ours	%	Heu	Ours	%	Heu	Ours	%
208	1424	1406	2.3	×	×		×	×	
218	1424	1284	9.8	1424	1406	2.3	×	×	
220	1424	1236	13.2	1424	1307	8.2	1424	1406	2.3
300	1424	595	58.2	1424	618	56.6	1424	678	52.4
350	1424	435	69.4	1424	452	68.2	1424	481	66.2
416	356	348	2.3	1424	430	69.0	1424	438	69.2
436	356	316	11.2	356	348	2.3	1424	424	70.2
440	356	303	14.8	356	335	6.8	356	348	2.3
500	356	236	33.7	356	262	26.4	356	266	25.3
550	356	148	58.4	356	154	56.7	356	164	54.0
600	356	98	72.5	356	98	72.5	356	98	72.5
Ave. Redu.(%)		55.5				58.7			61.3

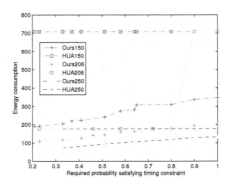

Figure 3.4. Guaranteed probability's impact on energy consumption.

We also give the impact of guaranteed probability on energy consumption in Figure 3.4. The experiment was implemented on Voltera Filter. The timing constraints are fixed. We selected three timing constraints: 150, 206, and 250. Figure 3.4 shows the energy consumption values from algorithm *HUA* and *VAP_S*. The three lines with square box stands for the results from algorithm *HUA*, and the three lines with "+" stands for the results from algorithm *VAP_S*. We can find that the energy consumption from algorithm *VAP_S* is significant lower than the results from *HUA*. In curve *HUA206*, there is a jump or energy consumption from 708 to 177 at probability 0.9. The reason is: at this time, we can switch the voltage level from high level $R1$ to $R2$. While at the probability below 0.9, we cannot switch according to algorithm *HUA*. The results from algorithm *VAP_S* can freely choose voltage level for each node and hence achieve impressive energy saving.

Figure 3.5 depicts timing constraints' impact on energy consumption. We fixed the guaranteed probability as 0.80. Figure 3.4 shows the energy consumption values from algorithm *HUA* and *VAP_S*. The solid line with "+" stands for the results from algorithm

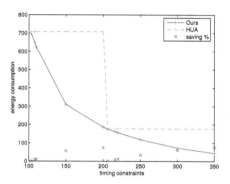

Figure 3.5. Timing constraints' impact on energy consumption.

VAP_S. It decreases quickly from 701 to 45 as the timing constraint increases from 103 to 350. The dashed line with square box stands for the results from algorithm *HUA*. It keeps 708 unchanged from timing constraint 103 to 205; Then at 206, it drops sharply to 177; After that it keeps 177 from timing constraint 206 to 350. The reason of the drop from 708 to 177 is because the voltage level switched from high voltage $R1$ to low voltage $R2$ for all the nodes. Before timing constraint 206, we can switch the voltage from $R1$ to low voltage $R2$ for all nodes since we cannot find way to guarantee the probability of satisfying timing constraint to reach 0.80 under low voltage $R2$. The dot dashed line with "·" is the result curve of energy saving percentage of algorithm *VAP_S* to *HUA*. The percentage first increases gradually; Then at the middle, it dropped to nearly 0; After that, it increases gradually again. We can find that the energy consumption from algorithm *VAP_S* is significant lower that the results from *HUA*. The larger the timing constraints, the lower the energy needed. This matches our expectation.

3.6 Conclusion

This chapter proposed two optimal algorithms to minimize energy in real-time DSP systems. By taking advantage of the uncertainties in execution time of tasks, our approach relaxes the rigid hardware requirements for software implementation and eventually avoids over-designing the system. For the *voltage assignment with probability* (VAP) problem, by using *Dynamic Voltage Scaling* (DVS), we proposed two algorithms, *VAP_S* and *VAP_M*, to give the optimal solutions for uniprocessor or multiprocessor DSP systems, respectively. Experimental results showed that our proposed algorithms achieved significant improvement on energy consumption savings than previous work. Our approach has great potential for further exploration.

CHAPTER 4

EFFICIENT ADAPTIVE ONLINE ENERGY-AWARE ALGORITHM FOR HETEROGENEOUS SENSOR NETWORKS

Energy and timing are critical issues for wireless sensor networks since most sensors are equipped with non-rechargeable batteries that have limited lifetime. However, sensor nodes usually work under dynamic changing and hard-to-predict environments. This chapter uses a novel *adaptive online energy-saving* (AOES) algorithm to save total energy consumption for heterogeneous sensor networks. Due to the uncertainties in execution time of some tasks and multiple working mode of each node, this chapter models each varied execution time as a probabilistic random variable to save energy by selecting the best mode assignment for each node, which is called MAP (Mode Assignment with Probability) problem. We propose an optimal sub-algorithm *MAP_Opt* to minimize the total energy consumption while satisfying the timing constraint with a guaranteed confidence probability. The experimental results show that our approach achieves significant energy saving than previous work.

4.1 Introduction

Recent advances in heterogeneous wireless communications and electronics have enabled the development of low cost, low power, multifunctional sensor nodes that are small in size and communicate in short distances. These tiny sensor nodes have capability to sense, process data, and communicate [1, 89]. Typically they are densely deployed in large numbers, prone to failures, and their topology changes frequently. They have limited power, compu-

103

tational capacity, bandwidth and memory. As a result of its properties, traditional protocols cannot be applied in this domain [18, 40, 47].

Sensor networks have a wide variety of applications in both military and civil environment. Some of these applications, e.g., natural habitat monitoring, require a large number of tiny sensors and these sensors usually operate on limited battery power. Individual sensors can last only 100-120 hours on a pair of AAA batteries in the active mode. On the other hand, since the number of sensors is huge and they may be deployed in remote, unattended, and hostile environments, it is usually difficult, if not impossible, to recharge or replace their batteries. This problem is compounded by the fact that battery capacity only doubles in 35 years. In a multi-hop ad hoc sensor network, each node plays the dual role of data originator and data router. The malfunctioning of a few nodes can cause significant topological changes and might require rerouting of packets and reorganization of the network [9, 13, 27]. It is for these reasons that researchers are currently focusing on the design of power-aware protocols and algorithms for sensor networks. The main task of a sensor node in a sensor field is to detect events, perform quick local data processing, and then transmit the data. Power consumption can hence be divided into three domains: sensing, communication, and data processing.

Lifetime of distributed micro sensor nodes is a very important issue in the design of sensor networks. The wireless sensor node, being a microelectronic device, can only be equipped with a limited power source ($\leq$ 0.5 Ah, 1.2 V). In some application scenarios, replenishment of power resources might be impossible. Hence, power conservation and power management take on additional importance [46, 102]. Optimal energy consumption, i.e., minimizing energy consumed by sensing and communication to extend the network lifetime, is an important design objective [84, 105]. To minimize energy consumption and extend network lifetime, a common technique is to put some sensors in the sleep mode and put the others in the active mode for the sensing and communication tasks. When a sensor is in the sleep mode, it is shut down except that a low-power timer is on to wake itself up at a later time [24, 25]. Therefore, it consumes only a tiny fraction of the energy consumed in the active mode [46, 83].

In the data transmission, real-time is a critical requirement for many application for wireless sensor network. There are three modes (active, vulnerable, and sleep) for a sensor network. We call it as *Mode Assignment with Probability* (MAP) problem. For example, in a Bio-sensor, we sample the temperature every minute. The data collected will go through a fixed topology to the destination. Assume we need the data transmission within 20 seconds. Given this requirement, we need to minimize the total energy consumed in each transmission. Due to the transmission line situation and other overheads, the execution time of each transmission is not a fix number. It may transmit a data in 1 seconds with 0.8 probability and in 3 seconds with 0.2 probability. The mode of a sensor node will affect both the energy and delay of the node.

Sensor networks usually work under dynamic changing and hard-to-predict environments. In this chapter, we propose an *adaptive online energy-saving* (AOES) algorithm to reduce the total energy consumption of heterogeneous sensor networks. First, we collect data at certain time interval. During each time interval, the execution time T of each node is estimated by an estimator. We mode T of each node as a random variable and predict the PDF (probability distribution function) of it [37, 38, 92, 111]. Then, we use our optimal *MAP_Opt* algorithm to solve the MAP problem. After finding the best mode assignment for each node, we use dynamic adaptive architecture to adjust the mode of each node online.

For heterogeneous systems [4], each node has different energy consumption rate, which related to area, size, reliability, etc. [26, 32, 42, 52]. Faster one has higher energy consumption while slower one has lower consumption. This chapter shows how to assign a proper mode to each node of a *Probability Data Flow Graph* (PDFG) such that the total energy consumption is minimized while the timing constraint is satisfied with a guaranteed confidence probability. With confidence probability P, we can guarantee that the total execution time of the PDFG is less than or equal to the timing constraint with a probability that is greater than or equal to P. In this chapter, we compare our optimal sub-algorithm *MAP_Opt* with a heuristic sub-algorithm *MAP_CP*, which is a revised version of previous work. Experiments show significant energy-saving improvement of our algorithm compared with *MAP_CP*.

Our contributions are listed as the following:

- Our algorithm can achieve significant energy-saving for heterogeneous distributed sensor network.

- Our algorithm *MAP_Opt* gives the optimal solution and achieves significant energy saving than *MAP_CP* algorithm.

- Our algorithm *MAP_Opt* not only is optimal, but also provides more choices of smaller total energy consumption with guaranteed confidence probabilities satisfying timing constraints. In many situations, algorithm *MAP_CP* cannot find a solution, while ours can find satisfied results.

- Our algorithm is practical and quick.

The rest of this chapter is organized as following: The models and basic concepts are introduced in Section 4.2. In Section 4.3, we give a motivational example. In Section 4.4, we propose our algorithms. The experimental results are shown in Section 4.5, and the conclusion is shown in Section 4.6.

4.2 System Model

In this section, we introduce some basic concepts and models which will be used in the later sections. First, we introduce the working model of senor networks. Next, we introduce the system model of PDFG. Finally, we define the MAP problem.

Working Model of Heterogeneous Sensor Networks:

Sensor usually works in highly continue changing environments. The work load and execution time of each node will change dynamically. We modeled the system dynamics with discrete events formulated over a fixed time interval or time window used to obtain the future performance requirements of the system.

For the adaptive architecture, the working procedures are as follows. During look ahead for the next time interval, The work load and corresponding execution time of each node is estimated. In each time interval, we find the best mode assignment to minimize total energy consumption while satisfying timing constraints with guaranteed probabilities. After finding the best assignment, the controller accordingly makes changes to the pool of hardware with the updated policy. The time interval is a design parameter, and will have to be decided by the designer based on empirical data obtained from simulations of the particular application.

System Model of PDFG:

Probabilistic Data-Flow Graph (PDFG) is used to model a sensor network application. A *PDFG G* $= \langle U, ED, T, M \rangle$ is a *directed acyclic graph* (DAG), where $U = \langle u_1, \cdots, u_i, \cdots, u_N \rangle$ is the set of nodes; $ED \subseteq U \times U$ is the edge set that defines the precedence relations among nodes in U. $M = \langle M_1, \cdots, M_j, \cdots, M_R \rangle$ is a mode set; the execution time $T_{M_j}(u)$ is a random variable; There is a timing constraint L and it must be satisfied for executing the whole PDFG.

In sensor network, we know that there are three kinds of mode, i.e., active, vulnerable, and sleep modes. The energy consumption is same when nodes are under same mode, and the execution time of each node is a random variable. We assume that from source to destination there is a fixed steps to go through. The Data Flow Graph is assumed to be a DAG (Directed Acyclic Graph), that is, there is no cycle in it.

In sensor networks, a node has different kinds of working modes. Assume there are maximum R different modes in a mode set M=$\{M_1, M_2, \cdots , M_R\}$. For each mode, there are maximum K execution time variations, although each node may have different number of modes and execution time variations. An assignment for a PDFG G is to assign a mode to each node. Define an *assignment A* to be a function from domain U to range M, where U is the node set and M is the mode set. For a node $u \in U$, $A(u)$ gives the selected mode of node u.

In a PDFG G, each varied execution time is modeled as a probabilistic random variable. $T_{M_j}(u)$ $(1 \leq j \leq R)$ represents the execution times of each node $u \in U$ for mode j, and $P_{M_j}(u)$ $(1 \leq j \leq R)$ represents the corresponding probability function. And $E_{M_j}(u)$ $(1 \leq j \leq R)$ is used to represent the energy consumption of each node $u \in U$ for mode j, which is a fixed value.

Definition:

We define the *MAP (Mode Assignment with Probability)* problem as follows: Given R different modes: $M_1, M_2, \cdots, M_R$, a PDFG $G = \langle U, ED \rangle$ with $T_{M_j}(v)$, $P_{M_j}(v)$, and $E_{M_j}(u)$ for each node $u \in U$ executed on each mode M_j, a timing constraint L and a confidence probability P, find the mode for each node in assignment A that gives the *minimum total energy consumption E with confidence probability P under timing constraint L*.

4.3 Motivational Example

In our model, under the same mode (M), the execution time (T) of a task is a random variable, which is usually due to condition instructions or operations that could have different
execution times for different inputs. The energy consumption (E) depends on the mode
M. Under different modes, a task has different energy consumptions. The execution time
of a node in active mode is less than that of it in vulnerable mode, and they both are less
than the execution time of it in sleep mode. The relations of energy consumption are just
the reverse. This chapter shows how to assign a proper mode to each node of a *Probabilistic Data-Flow Graph* (PDFG) such that the total energy consumption is minimized while
satisfying the timing constraint with a guaranteed confidence probability.

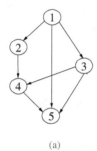

Nodes	M1			M2			M3		
	T	P	E	T	P	E	T	P	E
1	1	0.8	9	3	0.9	3	5	0.7	1
	2	0.2		4	0.1		6	0.3	
2	1	1.0	8	2	1.0	6	3	1.0	2
3	1	1.0	8	2	1.0	6	3	1.0	2
4	1	0.7	8	2	0.9	4	5	0.9	2
	3	0.3		4	0.1		6	0.1	
5	1	0.9	10	3	0.8	3	5	0.8	1
	2	0.1		4	0.2		6	0.2	

(a)   (b)

Figure 4.1. (a) A sensor network topology. (b) The times, probabilities, and energy consumptions of its nodes in different modes.

An exemplary PDFG is shown in Figure 4.1(a). Each node can select one of the
three different modes: M_1 (active), M_2 (vulnerable), and M_3 (sleep). The execution times
(T), corresponding probabilities (P), and energy consumption (E) of each node under
different modes are shown in Figure 4.1(b). The input DAG (*Directed Acyclic Graph*) has
five nodes. Node 1 is a multi-child node, which has three children: 2, 3, and 5. Node 5 is a
multi-parent node, and has three parents: 1, 3, and 4. The execution time T of each node

is modeled as a random variable. For example, when choosing M_1, node 1 will be finished in 1 time unit with probability 0.8 and will be finished in 2 time units with probability 0.2. Node 1 is the source and node 5 is the destination or the drain.

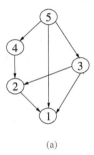

Nodes	M1			M2			M3		
	T	P	E	T	P	E	T	P	E
1	1	0.8	9	3	0.9	3	5	0.7	1
	2	0.2		4	0.1		6	0.3	
2	1	1.0	8	2	1.0	6	3	1.0	2
3	1	1.0	8	2	1.0	6	3	1.0	2
4	1	0.7	8	2	0.9	4	5	0.9	2
	3	0.3		4	0.1		6	0.1	
5	1	0.9	10	3	0.8	3	5	0.8	1
	2	0.1		4	0.2		6	0.2	

(a)   (b)

Figure 4.2. (a) The resulted DAG after topological sorting of Figure 4.1 (a). (b) The times, probabilities, and energy consumptions of its nodes in different modes.

The resulted DAG after topological sorting of Figure 4.1 (a) is shown in Figure 4.2 (a). And the times, probabilities, and energy consumptions of its nodes in different modes is shown in Figure 4.2 (b). The algorithm about how to implement this will be described in detail in later section.

In sensor network application, a real-time system does not always has hard deadline time. The execution time can be smaller than the hard deadline time with certain probabilities. So the hard deadline time is the worst-case of the varied smaller time cases. If we consider these time variations, we can achieve a better minimum energy consumption with satisfying confidence probabilities under timing constraints.

For Figure 4.1, the minimum total energy consumptions with computed confidence probabilities under the timing constraint are shown in Table 4.1. The results are generated by our algorithm, *MAP_Opt*. The entries with probability that is equal to 1 (see the entries in boldface) actually give the results to the hard real-time problem which shows the worst-case

Table 4.1. Minimum total energy consumptions with computed confidence probabilities under various timing constraints.

T	(P , E)	(P , E)	(P , E)	(P , E)	(P , E)
4	0.50, 43				
5	0.65, 39				
6	0.65, 35	0.81, 39			
7	0.65, 27	0.73, 33	0.81, 35	0.90, 39	
8	0.81, 27	0.90, 35	**1.00, 43**		
9	0.58, 20	0.73, 21	0.81, 27	0.90, 32	**1.00, 39**
10	0.72, 20	0.81, 21	0.90, 28	**1.00, 36**	
11	0.65, 14	0.90, 20	**1.00, 32**		
12	0.81, 14	0.90, 20	**1.00, 28**		
13	0.65, 12	0.90, 14	**1.00, 20**		
14	0.81, 12	0.90, 14	**1.00, 20**		
15	0.50, 10	0.90, 12	**1.00, 14**		
16	0.72, 10	0.90, 12	**1.00, 14**		
17	0.90, 10	**1.00, 12**			
18	0.50, 8	0.90, 10	**1.00, 12**		
19	0.72, 8	**1.00, 10**			
20	0.90, 8	**1.00, 10**			
21	**1.00, 8**				

scenario of the MAP problem. For each row of the table, the E in each (P, E) pair gives the minimum total energy consumption with confidence probability P under timing constraint j. For example, using our algorithm, at timing constraint 12, we can get (0.81, 14) pair. The assignments are shown in Table 4.2. We change the mode of nodes 2 and 3 to be M_2. Hence, we find the way to achieve minimum total energy consumption 14 with probability 0.81 satisfying timing constraint 12. While using the heuristic algorithm MAP_CP [80], the total energy consumption obtained is 28. Assignment $A(u)$ represents the voltage selection of each node u. There are 50% energy saving comparing with MAP_CP.

Table 4.2. The assignments of two algorithms *MAP_Opt* and *MAP_CP* with timing constraint 12.

		Node id	T	M	Prob.	Consum.
Ours	$A(u)$	1	3	M_2	0.90	3
		2	3	M_3	1.00	2
		3	3	M_3	1.00	2
		4	2	M_2	0.90	4
		5	4	M_2	1.00	3
	Total		12		0.81	14
MAP_CP	$A(u)$	1	2	M_1	1.00	9
		2	2	M_2	1.00	6
		3	2	M_2	1.00	6
		4	4	M_2	1.00	4
		5	4	M_2	1.00	3
	Total		12		1.00	28

4.4 The Algorithms For MAP Problem

In this section, we will propose our algorithms to solve the energy-saving problem of heterogeneous sensor networks. The basic idea is to use a time interval (window) and obtain the best mode assignment of each node in the time interval (window). Then adjust the mode of each node accordingly online. We proposed an algorithm, AOES (*Adaptive Online Energy-Saving Algorithm*), to reduce energy consumption while satisfying performance requirements for heterogeneous sensor networks. In this algorithm, we use *MAP_Opt* sub-algorithm to give the best mode assignment for each sensor node.

4.4.1 Adaptive Online Energy-Saving Algorithm

The AOES algorithm is shown in Algorithm 4.4.1. In AOES algorithm, we use the adaptive model to solve energy-saving problem for heterogeneous sensor networks. The adaptive approach includes three steps: First, collect updated information of each node and predict the PDF of execution time of each node in a time interval (window). Second, in

---
Algorithm 4.4.1 Adaptive Online Energy-Saving Algorithm *AOES*

Input: A sensor network, R different mode types, and the timing constraint L.

Output: a mode assignment to minimize energy E while satisfying L for the sensor network.

1: Collect data and predict the PDF of execution time of each node in a time interval (window).
2: Obtain the best mode assignment A by using *MAP_Opt* for each node in the time interval (window).
3: Output results: A and E_{min}.
4: Use online architectural adaptation to reduce energy consumption while satisfying timing constraints with guaranteed probability.
5: Repeat the above steps.

---

each time interval (window), obtain the best mode assignment for each node during the time interval (window) to minimize the energy consumption while satisfying timing constraint with guaranteed probability. Third, use an on-line architecture adaptation control policy. Since our design is for a non-stationary environments, the control policy varies with the environment but is stationary within a time interval (window).

4.4.2 The MAP_CP Algorithm

In this subsection, we first design an heuristic algorithm for sensor network. We call this algorithm as *MAP_CP*.

The *MAP_CP* algorithm is shown in Algorithm 4.4.2. A *critical path* (CP) of a DAG is a path from source to its destination. To be a legal assignment for a PDFG, the execution time for any critical path should be less than or equal to the given timing constraint. In algorithm *MAP_CP*, we only consider the hard execution time of each node, that is, the case when the probability of the random variable T equals 1. This is a heuristic solution for hard real-time systems. We find the CP with minimized energy consumption first, then adjust the energy of the nodes in CP until the total execution time is less than or equal to L.

Algorithm 4.4.2 Heuristic algorithm for the MAP problem when the PDFG is DAG (*MAP_CP*)

Input: R different mode types, a DAG, and the timing constraint L.

Output: a mode assignment to minimize energy while satisfying L.

1: Assign the lowest energy type to each node and mark the type as assigned.
2: Find a CP that has the maximum execution time among all possible paths based on the current assigned types for the DAG.
3: For every node u_i in CP,
4:     for every unmarked type p,
5:         change its type to p,
6:         calculate $r = cost\_increase/time\_reduce$
7:         select the minimum r.
8:         if $(T > L)$
9:             contiune
10:        else
11:            exit   /* This is the best assignment */

4.4.3   The MAP_Opt Algorithm

For sensor network, we propose our algorithm, *MAP_Opt*, which is shown as follows.

Input: R different modes, a DAG, and the timing constraint L.

Output: An optimal mode assignment

1. Topological sort all the nodes, and get a sequence A.

2. Count the number of multi-parent nodes t_{mp} and the number of multi-child nodes t_{mc}. If $t_{mp} < t_{mc}$, use bottom up approach; Otherwise, use top down approach.

3. For bottom up approach, use the following algorithm. For top down approach, just reverse the sequence. $|V| \leftarrow N$, where $|V|$ is the number of nodes.

4. If the total number of nodes with multi-parent is t, and there are maximum K variations for the execution times of all nodes, then we will give each of these t nodes a fixed assignment.

5. For each of the K^t possible fixed assignments, assume the sequence after topological sorting is $u_1 \rightarrow u_2 \rightarrow \cdots \rightarrow u_N$, in bottom up fashion. Let $D_{1,j} = B_{1,j}$. Assume $D'_{i,j}$ is the table that stored minimum total energy consumption with computed confidence probabilities under the timing constraint j for the sub-graph rooted on u_i except u_i. Nodes $u_{i_1}, u_{i_2}, \cdots, u_{i_w}$ are all child nodes of node u_i and w is the number of child nodes of node u_i, then

$$D'_{i,j} = \begin{cases} (0,0) & \text{if } w = 0 \\ D_{i_1,j} & \text{if } w = 1 \\ D_{i_1,j} \oplus \cdots \oplus D_{i_w,j} & \text{if } w \geq 1 \end{cases} \tag{4.1}$$

6. Then, for each k in $B_{i,k}$.

$$D_{i,j} = D'_{i,j-k} \oplus B_{i,k} \tag{4.2}$$

7. For each possible fixed assignment, we get a $D_{N,j}$. Merge the (Probability, Energy) pairs in all the possible $D_{N,j}$ together, and sort them in ascending sequence according probability.

8. Then remove redundant pairs. Finally get $D_{N,j}$.

In algorithm *MAP_Opt*, we exhaust all the possible assignments of multi-parent or multi-child nodes. Without loss of generality, assume we using bottom up approach. If the total number of nodes with multi-parent is t, and there are maximum K variations for the execution times of all nodes, then we will give each of these t nodes a fixed assignment. We will exhausted all of the K^t possible fixed assignments. Algorithm *MAP_Opt* gives the optimal solution when the given PDFG is a DAG. In equation (4.1), $D_{i_1,j} \oplus D_{i_2,j}$ is computed as follows. let G' be the union of all nodes in the graphs rooted at nodes u_{i_1} and u_{i_2}. Travel all the graphs rooted at nodes u_{i_1} and u_{i_2}. For each node a in G', we add the energy consumption of a and multiply the probability of a to $D'_{i,j}$ for only once, because each node can only have one assignment and there is no assignment conflict. The final

$D_{N,j}$ we get is the table in which each entry has the minimum energy consumption with a guaranteed confidence probability under the timing constraint j.

In algorithm *MAP_Opt*, there are K^t loops and each loop needs $O(|V|^2 * L * R * K)$ running time. The complexity of *Algorithm MAP_Opt* is $O(K^{t+1} * |V|^2 * L * R)$. Since t_{mp} is the number of nodes with multi-parent, and t_{mc} is the number of nodes with multi-child, then $t = min(t_{mp}, t_{mc})$. $|V|$ is the number of nodes, L is the given timing constraint, R is the maximum number of modes for each node, and K is the maximum number of execution time variation for each node. The experiments show that algorithm *MAP_Opt* runs efficiently.

4.5 Experiments

This section presents the experimental results of our algorithms. We conduct experiments on a set of DAGs. Three different modes, M_1, M_2, and M_3, are used in the system, in which a node with mode M_1 (active) is the quickest with the highest energy consumption and a node with type M_3 (sleep) is the slowest with the lowest energy consumption. The distribution of execution times of each node is Gaussian.

We compare two methods base on algorithms *MAP_CP* and *MAP_Opt*.

Method 1: Algorithm *AOES* with sub-algorithm *MAP_CP*.

Method 2: Algorithm *AOES* with sub-algorithm *MAP_Opt*.

The experiments are performed on a Dell PC with a P4 2.1 G processor and 512 MB memory running Red Hat Linux 9.

Figure 4.3 shows a DAG with 17 nodes. This is for exp1 in our six experiments. We assume this is the topology of a sensor network. S is the source and D is the destination. Each node has three modes with different execution times and energy consumptions. The

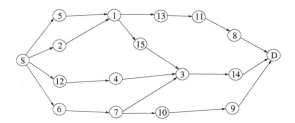

Figure 4.3. The PDFG of exp1.

collected data need to go through the topology to the destination within a timing constraint. Exclude the source and destination node, this DAG has 3 multi-child nodes and 5 multi-parent nodes. Using top-down approach, we implemented all $3^3 = 27$ possibilities. The experimental results for exp1 is shown in Table 4.3. Column "TC" stands for the timing constraint of the DAG. Column "Saving(% 1)" shows the percentage of reduction on system energy consumptions, compared the results of soft real-time with those of hard real-time, in other words, compared the results of Method 2 with those of Method 1. The average percentage reduction is shown in the last row "Average Saving" of the table. The entry with "×" means no solution available. Under timing constraint 50 in Table 4.3, there is no solution for hard real-time for Method 2 using *MAP_CP* algorithm. However, we can find solution 6138 with probability 0.9 that guarantees the total execution time of the DFG is less than or equal to the timing constraint 50.

The experimental results of all our six experiments are shown in Table 4.4. "Exp" is the name of experiments, and "N" stands for the number of nodes of a DAG. Table 4.4 shows that our algorithms can greatly reduce the total energy consumption while have a guaranteed confidence probability satisfying timing constraints. On average, Method 1 gives an energy reduction of 33.3% with confidence probability 0.9 under timing constraints, and an energy reduction of 41.4% and 48.0% with 0.8 and 0.7 confidence prob-

Table 4.3. Experimental results of Method 1 and Method 2 for exp1.

| TC | Method 1 Energy | Method 2 | | | | | |
| | | 0.7 | | 0.8 | | 0.9 | |
		Energy	Saving(% 1)	Energy	Saving(% 1)	Energy	Saving(% 1)
50	×	6138		×		×	
60	×	6124		6125		6178	
70	×	5570		5575		6173	
80	**6223**	4502	27.7%	4992	19.8%	4754	23.6%
90	**6216**	3307	46.8%	3890	37.4%	4267	31.3%
100	**2874**	1302	54.7%	1890	34.2%	2465	14.2%
120	**3735**	1302	65.1%	1886	49.5%	1795	51.9%
140	**3610**	1302	63.9%	1302	63.9%	1795	50.3%
165	**1807**	1302	27.9%	1302	27.9%	1239	31.4%
166	**1302**	1302		1302		1302	
Average Saving			47.7%		38.8 %		33.8 %

abilities satisfying timing constraints, respectively. The experiments using *MAP_Opt* on these DAGs are finished within several minutes.

Table 4.4. Experimental results of algorithms Method 1 and Method 2 for different DAGs

| Exp | N | Method 1 Energy | Method 2 | | | | | |
| | | | 0.7 | | 0.8 | | 0.9 | |
			Energy	Saving(% 1)	Energy	Saving(% 1)	Energy	Saving(% 1)
exp1	24	2982	1558	47.8%	1850	38.0%	2008	32.5%
exp2	35	4215	2283	45.8%	2378	43.6%	2687	36.3%
exp3	43	5176	2572	50.3%	2953	43.0%	3513	32.1%
exp4	62	6402	3348	47.7%	3812	40.5%	4410	31.1%
exp5	78	9736	5073	47.9%	5736	41.1%	6427	34.0%
exp6	89	11865	6120	48.4%	6815	42.6%	7827	33.9%
Average Saving				48.0%		41.4%		33.3%

The advantages of Method 2 over Method 1 are summarized as follows. First, our algorithms are efficient and provide overview of all possible variations of minimum costs comparing with the the worst-case scenario generated by the *MAP_CP*. Although using probabilistic approach the MAP problem becomes very complicate, our algorithms give very good results. More information and choices are provided by our algorithms. Second, it is possible to greatly reduce the system total energy consumption while have a very high confidence probability under different timing constraints. Third, given an assignment, we are able to get the minimum total energy consumption with different confidence probabilities under each timing constraint. Finally, our algorithms are very quick and practical.

4.6 Conclusion

This chapter proposed a novel *adaptive online energy-saving* (AOES) algorithm to save total energy consumption for heterogeneous sensor networks. Due to the uncertainties in execution time of some tasks and multiple working mode of each node, we modeled each varied execution time as a probabilistic random variable to save energy by selecting the best mode assignment for each node, which is called MAP (Mode Assignment with Probability) problem. We proposed an optimal sub-algorithm *MAP_Opt* to minimize the total energy consumption while satisfying the timing constraint with a guaranteed confidence probability. The experimental results demonstrated the effectiveness of our method.

CHAPTER 5

ENERGY MINIMIZATION WITH GUARANTEED PROBABILITY SATISFYING TIMING CONSTRAINT VIA DVS AND LOOP SCHEDULING

Low energy consumption is an important problem in real-time embedded systems and loop is the most energy consuming part in most cases. As in Chapter 3, this chapter models each varied execution time as a probabilistic random variable. We use rotation scheduling and DVS (Dynamic Voltage Scaling) to minimize the expected total energy consumption while satisfying the timing constraint with a guaranteed confidence probability. Our approach can handle loops efficiently. In addition, it is suitable to both soft and hard real-time systems. And even for hard real-time, we have good results. The experimental results show that our approach achieves significant energy saving than voltage assignment using list scheduling and ILP (Integer Linear Programming). For example, our algorithm achieves an average improvement of 32.6% on total energy consumption reduction over ILP for hard real-time.

5.1 Introduction

Energy consumption has become a primary concern in today's real-time embedded systems. [101]. In DSP systems, some tasks may not have fixed execution time. Such tasks usually contain conditional instructions and/or operations that could have different execution times for different inputs [37, 92]. It is possible to obtain the execution time distribution for each task by sampling or profiling [91]. Prior design space exploration methods for hardware/software codesign of embedded systems [42, 43, 65, 80] guarantee no deadline missing by considering worst-case execution time of each task. These methods are pessimistic and will often lead to over-designed systems with high cost. There are several papers on the probabilistic timing performance estimation for soft real-time systems design [37]. They modeled each task's execution time as a random variable.

In this chapter, we use probabilistic approach and loop scheduling to avoid over-design systems. We propose a novel optimal algorithm to minimize expected value of total energy consumption while satisfying timing constraints with guaranteed probabilities for real-time applications.

Dynamic voltage scaling (DVS) is one of the most effective techniques to reduce energy consumption [20, 79, 82, 110]. Many researches have been done on DVS for real-time applications in recent years [79, 82, 110]. Zhang et. al. [110] proposed an ILP (Integer Linear Programming) model to solve DVS on multiple processor systems. Shin et. al. [82] proposed a DVS technique for real-time applications based on static timing analysis. However, in the above work, applications are modeled as DAG (Directed Acyclic Graph), and loop optimization is not considered. Saputra et. al. [79] considered loop optimization with DVS. However, in their work, the whole loop is scaled with the same voltage. Our technique can choose the best voltage level assignment for each task node.

We design new rotation scheduling algorithms for real-time applications that produce schedules consuming minimal energy. In our algorithms, we use rotation scheduling [15, 17] to get schedules for loop applications. The schedule length will be reduced after rotation. Then, we use DVS to assign voltages to computations individually in order to decrease the voltages of processors as much as possible within the timing constraint. The experimental data show that our algorithms can get better results on energy saving than the previous work. On average, VASP_RS shows a 32.6% reduction in energy consumption compared with the ILP technique in [110] for hard real-time. For soft real-time embedded systems, the experimental results are even better.

The rest of this chapter is organized as following: In Section 5.2, we give motivational examples. The models and basic concepts are introduced in Section 5.3. In Section 5.4, we propose our algorithms, *VASP_RS*. The experimental results are shown in Section 5.5, and the conclusion is shown in Section 5.6.

5.2 Motivational Examples

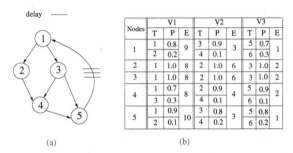

Nodes	V1			V2			V3		
	T	P	E	T	P	E	T	P	E
1	1	0.8	9	3	0.9	3	5	0.7	1
	2	0.2		4	0.1		6	0.3	
2	1	1.0	8	2	1.0	6	3	1.0	2
3	1	1.0	8	2	1.0	6	3	1.0	2
4	1	0.7	8	2	0.9	4	5	0.9	2
	3	0.3		4	0.1		6	0.1	
5	1	0.9	10	3	0.8	3	5	0.8	1
	2	0.1		4	0.2		6	0.2	

(a)　　　　　　　　　　　　(b)

Figure 5.1. (a) A PDFG. (b) The times, probabilities, and energy consumptions of its nodes under different voltage levels.

Assume an input PDFG (*Probability Data Flow Graph*) shown in Figure 5.1(a). Each node can select one of the three different voltages: V_1, V_2, and V_3. The execution times (T), corresponding probabilities (P), and expected energy consumption (E) of each node under different voltage levels are shown in Figure 5.1(b). The input PDFG has five nodes. Node 1 is a multi-child node, which has two children: 2 and 3. Node 5 is a multi-parent node, and has two parents: 3 and 4. The execution time T of each node is modeled as a random variable. For example, When choosing V_1, node 1 will be finished in 1 time unit with probability 0.8 and will be finished in 2 time units with probability 0.2.

Figure 5.2(b) shows the static schedule of (a), that is, the DAG without delay edge. Figure 5.2(c) shows the schedule graph of (b) using list scheduling.

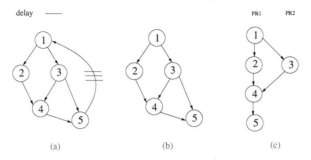

Figure 5.2. (a) Original PDFG. (b) The static schedule (DAG). (c) The schedule graph using list scheduling.

For schedule graph in Figure 5.2 (b), the minimum total energy consumptions with computed confidence probabilities under the timing constraints are shown in Table 5.1. The results are generated by our algorithm, *VAP_M*, which is a sub-algorithm of our *VASP_RS* algorithm. The entries with probability that is equal to 1 (see the entries in boldface) actually give the results to the hard real-time problem which shows the worst-case scenario. For each row of the table, the E in each (P, E) pair gives the minimum total energy consumption with confidence probability P under timing constraint j. For example, using our algorithm, at timing constraint 11, we can get (0.90, 20) pair. The assignments are shown as "Ass_1" in Table 5.2. Assignment $A(v)$ represents the voltage selection of each node v. Hence, we find the way to achieve minimum total energy consumption 20 with probability 0.90 satisfying timing constraint 11. While using the ILP and heuristic algorithm in [80], the total energy consumption obtained is 32. The assignments are shown as "Ass_2" in Table 5.2.

Figure 5.3 (a) shows the assignment template with timing constraint 11 before rotation. This template gives us the idea of the relative positions of nodes in schedule graph. Figure 5.3 (b) shows the new relative positions of nodes after rotating down node 1. In

Table 5.1. Minimum total energy consumptions with computed confidence probabilities under various timing constraints.

T	(P , E)	(P , E)	(P , E)	(P , E)	(P , E)
4	0.50, 43				
5	0.65, 39				
6	0.65, 35	0.81, 39			
7	0.65, 27	0.73, 33	0.81, 35	0.90, 39	
8	0.81, 27	0.90, 35	**1.00, 43**		
9	0.58, 20	0.73, 21	0.81, 27	0.90, 32	**1.00, 39**
10	0.72, 20	0.81, 21	0.90, 28	**1.00, 36**	
11	0.65, 14	0.90, 20	**1.00, 32**		
12	0.81, 14	0.90, 20	**1.00, 28**		
13	0.65, 12	0.90, 14	**1.00, 20**		
14	0.81, 12	0.90, 14	**1.00, 20**		
15	0.50, 10	0.90, 12	**1.00, 14**		
16	0.72, 10	0.90, 12	**1.00, 14**		
17	0.90, 10	**1.00, 12**			
18	0.50, 8	0.90, 10	**1.00, 12**		
19	0.72, 8	**1.00, 10**			
20	0.90, 8	**1.00, 10**			
21	**1.00, 8**				

Figure 5.4, (a) shows the retimed PDFG. (b) is the static schedule of (a). (c) is the graph after rotation. Table 5.3 shows the minimum total energy consumptions with confidence probabilities under different timing constraints for the new schedule graph. At timing constraint 11, we get (0.90, 10) and (1.00, 12) pairs. The detail assignment of each node is shown in Table 5.4. For (0.90, 10) pair, node 5's type was changed to V_3, then the T was changed from 4 to 6, and energy consumption change from 3 to 1. Hence the total execution time is 11, and the total energy consumption is 10. So the improvement of energy saving is 50.0% while the probability is still 90%. For (1.00, 12) pair, the execution times of nodes 2, 3 were changed from 1 to be 3 and node 1's was changed from 2 to 6. The total energy consumptions were changed to 12, and the total execution time is still 11. Hence compared with original 32, the improvement of total energy saving is 62.5%. If we consider the energy consumptions of switch activity, we can get more practical results. For example, after rotation once, node 1 has changed from processor $PR1$ to $PR2$. Assume

Table 5.2. The assignments with timing constraint 11.

		Node id	T	V Level	Prob.	Energy
		1	2	V_1	1.00	9
Ass_1	$A(u)$	2	3	V_3	1.00	2
		3	3	V_3	1.00	2
		4	2	V_2	0.90	4
		5	4	V_2	1.00	3
	Total		11		0.90	20
		1	2	V_1	1.00	9
Ass_2	$A(u)$	2	1	V_1	1.00	8
		3	1	V_1	1.00	8
		4	4	V_2	1.00	4
		5	4	V_2	1.00	3
	Total		11		1.00	32

the energy consumption of this switch is 1, then the final total energy consumption is 13. The energy saving is 59.4% compared with previous scheduling and assignment.

5.3   Models and Concepts

In this section, we introduce the energy model, the system model, and the VASP problem. Since the energy model and system model have been introduced in Chapter 3, this chapter will emphasize on the VASP problem.

Static Schedules: From the PDFG of an application, we can obtain a static schedule. A static schedule [16] of a cyclic PDFG is a repeated pattern of an execution of the corresponding loop. In our works, a schedule implies both control step assignment and allocation. A static schedule must obey the dependency relations of the Directed Acyclic Graph (DAG) portion of the PDFG. The DAG is obtained by removing all edges with delays in the PDFG.

Retiming: Retiming [49] is an optimal scheduling technique for cyclic PDFGs considering inter-iteration dependencies. It can be used to optimize the cycle period of a cyclic PDFG by evenly distributing the delays. Retiming generates the optimal schedule

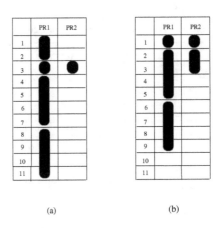

(a)   (b)

Figure 5.3. (a) The template for assignment (b) The template for rotation.

for a cyclic PDFG when there is no resource constraint. Given a cyclic PDFG G=⟨U, ED, d, T, V⟩, retiming r of G is a function from U to integers. For a node $u \in U$, the value of r(u) is the number of delays drawn from each of incoming edges of node u and pushed to all of the outgoing edges. Let G_r=⟨U, ED, d_r, T, V⟩ denote the retimed graph of G with retiming r, then $d_r(ed) = d(ed) + r(u) - r(v)$ for every edge $ed(u \rightarrow v) \in$ ED.

Rotation Scheduling: Rotation Scheduling [15] is a scheduling technique used to optimize a loop schedule with resource constraints. It transforms a schedule to a more compact one iteratively in a PDFG. In most cases, the minimal schedule length can be obtained in polynomial time by rotation scheduling. Figure 5.5 shows an example to explain how to obtain a new schedule via rotation scheduling. We use the schedule generated by list scheduling in Figure 5.1(a) as an initial schedule. In Figure 5.5(a) we change the node label 1, 2, 3, 4, 5 to A, B, C, D, E. Figure 5.5(c) shows the corresponding code. We get a set of nodes at the first row of the schedule, in this case, it is {1}, and we rotate node 1

delay ——

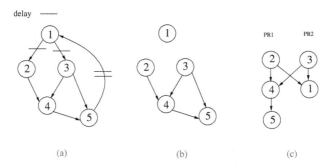

(a)   (b)   (c)

Figure 5.4. (a) The PDFG after retiming (b) The static schedule. (c) The schedule after rotation.

down. The rotated graph is shown in Figure 5.5(b). The equivalent loop body after rotation is shown in Figure 5.5(d). The code size is increased by introducing the prologue and epilogue after the rotation is performed, which can be solved by the technique proposed in [113].

Definitions: Define the *VASP (voltage assignment and scheduling with probability)* problem as follows: Given M different voltage levels: $V_1, V_2, \cdots, V_M$, a PDFG $G = \langle U, ED, T, V \rangle$ with $T_{V_j}(u)$, $P_{V_j}(u)$, and $E_{V_j}(u)$ for each node $u \in U$ executed on each voltage V_j, a timing constraint L and a confidence probability P, find the voltage for each node in assignment A that gives the *minimum expected total energy consumption E with confidence probability P under timing constraint L*.

5.4   The Algorithms with Probabilistic Design Space Exploration and DVS

In this section, we propose a novel algorithm, *VASP_RS*, to solve the VASP problem. The basic idea is to use rotation scheduling and DVS to repeatedly regroup a loop and decrease the energy by voltage selection as much as possible within a timing constraint. In algorithm *VASP_RS*, we need use three sub algorithms: *VAP_SG*, *VAP_M*, and *RS*. Algorithms

Table 5.3. Minimum total energy consumptions with computed confidence probabilities under various timing constraints.

T	(P , E)	(P , E)	(P , E)	(P , E)
3	0.63, 43			
4	0.73, 33	0.81, 39		
5	0.73, 29	0.90, 33		
6	0.73, 21	0.90, 29	**1.00, 37**	
7	0.50, 20	0.90, 24	**1.00, 31**	
8	0.50, 12	0.72, 14	0.90, 20	**1.00, 28**
9	0.90, 12	**1.00, 24**		
10	0.72, 10	0.90, 12	**1.00, 20**	
11	0.90, 10	**1.00, 12**		
12	0.90, 10	**1.00, 12**		
13	0.72, 8	**1.00, 10**		
14	0.90, 8	**1.00, 10**		
15	**1.00, 8**			

VAP_SG and *VAP_M* have already been described in detail in Chapter 3. Hence here we only describe the *RS* algorithm.

The VASP_RS Algorithm

Algorithm 5.4.1 Optimal algorithm for the VASP problem (***VASP_RS***)

Input: M different voltage levels, a PDFG, the timing constraint L, and the rotation times R.

Output: An optimal voltage assignment to minimize energy while satisfying L

1: rotation $\leftarrow$ 0;
2: while (rotation $< R$)
3:     Get the static schedule (a DAG) from input PDFG.
4:     Get scheduling graph from the DAG using algorithm *VAP_SG*.
5:     Using algorithm *VAP_M* to get the near optimal assignment of voltage levels for the schedule graph.
6:     Using Rotation scheduling *RS* to retime the original PDFG and rotate down the first row of template.
7:     rotation++;

The RS Algorithm

We show the rotation scheduling algorithm with our PDFG model. The main idea of this algorithms is: Firstly, get a shorter schedule by rotating an original schedule. Then using

Table 5.4. The assignments with timing constraint 11.

		Node id	T	V	Prob.	Energy
		1	6	V_3	1.00	1
Ass_1	$A(v)$	2	3	V_3	1.00	2
		3	3	V_3	1.00	2
		4	2	V_2	0.90	4
		5	6	V_3	1.00	1
	Total		11		0.90	10
		1	6	V_3	1.00	1
Ass_2	$A(v)$	2	3	V_3	1.00	2
		3	3	V_3	1.00	2
		4	4	V_2	1.00	4
		5	4	V_2	1.00	3
	Total		11		1.00	12

VAP_M to minimize energy consumption via better voltage level selection.

Algorithm 5.4.2 Rotation scheduling for the VASP problem (*RS*)

Input: M different voltage levels, a PDFG, and the timing constraint L.
Output: a voltage assignment to minimize energy while satisfying L.
1: Input the PDFG G and environment parameter.
2: Create a schedule S using list scheduling.
3: For i = 1 to n,
4:     Get a set of nodes U_1 to be rotated;
5:     Delete nodes in U_1 from S;
6:     For each $u \in S$, Retime v, and get new graph G_r;
7:     Compact S according to G_r;
8:     Insert nodes in U_1 into S according to G_r;
9: Output a new graph.

The *RS* algorithm is described in detail as following: first input a PDFG G and the environment parameters such as the number of function-units, the timing constraint, the repeat number of rotation n, the levels of voltages and the values of voltages. Then create a schedule table S in step 2 by using list scheduling. From step 3 to 9, there is a loop. In each loop body, we rotate the schedule and retime the graph. In step 4, get a set of nodes U_1 to be rotated, which are the nodes in the first row of S, and remove them from S. In step 6, modify the retiming function r of G when retime the nodes in U_1, and get a new PDFG

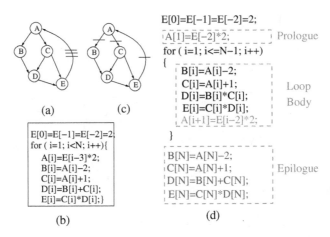

(a)   (c)

```
E[0]=E[-1]=E[-2]=2;
for ( i=1; i<N; i++){
  A[i]=E[i-3]*2;
  B[i]=A[i]-2;
  C[i]=A[i]+1;
  D[i]=B[i]+C[i];
  E[i]=C[i]*D[i];}
```

(b)

```
E[0]=E[-1]=E[-2]=2;
A[1]=E[-2]*2;                     Prologue
for ( i=1; i<=N-1; i++)
{
    B[i]=A[i]-2;
    C[i]=A[i]+1;                  Loop
    D[i]=B[i]*C[i];              Body
    E[i]=C[i]*D[i];
    A[i+1]=E[i-2]*2;
}
B[N]=A[N]-2;
C[N]=A[N]+1;                      Epilogue
D[N]=B[N]+C[N];
E[N]=C[N]*D[N];
```

(d)

Figure 5.5. (a) The original PDFG. (b) The rotated PDFG. (c) The code of loop application corresponding to (a). (d) The equivalent loop after regrouping loop body.

G_r. In step 7, compact S, which means find an earliest position for each node in S. Then in step 8 put the nodes in U_1 back to S according to the principle of As Early As Possible (AEAP),

5.5 Experiments

This section presents the experimental results of our algorithms. We compare our algorithm with list scheduling and the ILP techniques in [110] that can give near-optimal solution for the DAG optimization. We conduct experiments on a set of benchmarks including 4-stage lattice filter, 8-stage lattice filter, voltera filter, differential equation solver, RLS-languerre lattice filter, and elliptic filter. Three different voltage levels, V_1 (2.5), V_2 (1.8), and V_3 (1.2), are used in the system, in which a processor under V_1 is the quickest with the highest energy consumption and a processor under V_3 is the slowest with the lowest energy consumption. The distribution of execution times of each node is Gaussian. Each application has timing constraints. Let the effective switched capacitance $C_s = 1$ when the energy is calculated. This assumption is good enough to serve our purpose to compare the relative improvement among different algorithms. For each benchmark, we conduct two sets of the experiments based on 3 processor cores and 5 processor cores, respectively. The experiments are performed on a Dell PC with a P4 2.1 G processor and 512 MB memory running Red Hat Linux 9.0.

Table 5.5. The energy comparison for the schedules generated by list scheduling, the ILP in [110], and the VASP_SR algorithm.

TC	List	ILP	VASP_SR						
			0.8		0.9		1.0		
	Ene.	Ene.	Ene.	%I	Ene.	%I	Ene.	%I	
	(μJ)	(μJ)	(μJ)	(%)	(μJ)	(%)	(μJ)	(%)	
RLS-laguerre Filter, 3 processors									
60	1518	1265	541	57.2	688	45.6	841	33.5	
80	1325	993	461	53.6	565	43.1	677	31.8	
100	1231	876	448	48.8	532	39.2	624	28.8	
110	1128	733	314	57.1	394	46.3	461	37.1	
120	1052	710	313	55.9	378	46.7	451	36.5	
8-stage Lattice Filter, 3 processors									
160	3208	2309	1106	52.1	1312	43.2	1584	31.4	
180	2501	1625	736	54.7	902	44.5	1074	33.9	
200	2106	1495	728	51.3	855	42.8	1009	32.5	
220	1892	1381	695	49.7	831	39.8	941	31.2	
240	1587	1095	541	50.6	624	43.1	733	33.1	
Elliptic Filter, 3 processors									
100	2507	1830	626	54.6	963	47.4	1245	32.1	
120	2210	1658	746	55.1	857	48.3	1071	35.4	
140	1992	1395	653	53.2	765	45.2	927	33.5	
160	1880	1341	551	58.9	685	48.9	896	33.2	
170	1677	1209	510	57.8	640	47.1	785	35.1	
Average Imp. over ILP			–	55.6	–	46.8	–	32.6	

The experimental results with 3 processors are shown in Table 5.5. In the table, columns "List", "ILP", "VASP_RS" represent the results obtained by list scheduling, the

ILP in [110], our *VASP_RS* algorithm, respectively. Columns "Ene." represents the energy consumption and columns "%I" represents the improvement of our algorithm over the ILP in [110] with different probability. The total average improvement of *VASP_RS* is shown in the last row.

From the experimental results, we can see that our algorithms achieve significant energy saving compared with the ILP in [110]. On average, *VASP_RS* shows a 32.6% reduction in hard real-time, and reductions of 46.8%, 55.6% with probability 0.9 and 0.8, respectively, for soft real-time DSP systems. The reason of such big improvement is because we use loop scheduling and DVS to shrink the schedule length and assign the best possible voltage level to minimize energy while satisfying time constraint with guaranteed probabilities.

5.6 Conclusion

This paper combined loop scheduling and DVS to solve the VASP (*Voltage Assignment and Scheduling with Probability*) problem of real-time embedded systems. By taking advantage of the uncertainties in execution time of tasks, our approach give out voltage assignments and scheduling to minimize the expected total energy consumption while satisfying timing constraint with guaranteed probabilities. Our approach can eventually avoid over-designing real-time embedded systems. We repeatedly regroup a loop based on rotation scheduling and decrease the energy by voltage selection as much as possible within a timing constraint. We proposed an algorithm, *VASP_RS*, to give the efficient solutions. Experimental results show that our proposed algorithms achieved significant improvement on energy consumption savings than previous work. For both soft and hard real-time systems, the energy consumption reduction is significant. Furthermore, our approach can handle loops efficiently.

CHAPTER 6

LOOP SCHEDULING TO MINIMIZE COST WITH DATA MINING AND PREFETCHING FOR HETEROGENEOUS DSP

In real-time embedded systems, such as multimedia and video applications, cost and time are the most important issues and loop is the most critical part. We proposes a novel algorithm to minimize the total cost while satisfying the timing constraint with a guaranteed confidence probability. First, we use data mining to predict the distribution of execution time and find the association rules between execution time and different inputs from history table. Then we use rotation scheduling to obtain the best assignment for total cost minimization, which is called the HAP problem. Finally, we use prefetching to prepare data in advance at run time. Experiments demonstrate the effectiveness of our algorithm. Our approach can handle loops efficiently. In addition, it is suitable to both soft and hard real-time systems.

6.1 Introduction

In high level synthesis, cost (such as energy, reliability, etc.) minimization has become a primary concern in today's real-time embedded systems. In DSP systems, some tasks may not have fixed execution time. Such tasks usually contain conditional instructions and/or operations that could have different execution times for different inputs. It is possible to obtain the execution time distribution for each task by sampling or profiling [91]. In this chapter, we use a data mining engine to do the prediction. First, the data mining engine collects data into the log. Then it does clustering and uses unsupervising method to find

133

the distribution pattern of all random variables, i.e., the execution times. Finally, the engine builds the distribution function for each execution time that has uncertainty.

In heterogeneous DSP, i.e., there are multiple functional unit (FUs) type for each task to choose from. Different FU type has different execution time and cost, which may relate to energy, area, etc. A certain FU type may execute the task slower but with less cost, while another type will execute faster with higher cost. Prior design space exploration methods for hardware/software codesign of embedded systems [80] guarantee no deadline missing by considering worst-case execution time of each task. These methods are pessimistic and will often lead to over-designed systems with high cost. In this chapter, we use probabilistic approach and loop scheduling to avoid over-designing systems. We compute the best type assignment at compile time to minimize expected value of total energy consumption while satisfying timing constraints with guaranteed probabilities for real-time applications.

We design new rotation scheduling algorithms for real-time applications that produce schedules consuming minimal energy. In our algorithms, we use rotation scheduling [15] to get schedules for loop applications.

The schedule length will be reduced after rotation. Then, we assign different FU to computations individually in order to decrease the cost of processors as much as possible within the timing constraint.

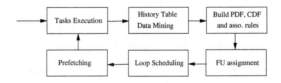

Figure 6.1. The basic implementation steps of our method

In summary, our approach includes three steps, which are shown in Figure 6.1. First, we use data mining to predict the execution time pattern and time-input association rules from history table. Second, we use rotation scheduling to obtain the best assignment for total cost minimization. This includes finding the best FU type assignment in each iteration and rotation scheduling for Q iterations. Finally, we use prefetching to prepare data in advance at run time. The experimental data show that our algorithms can get better results on cost saving than the previous work.

The rest of the chapter is organized as following: In Section 6.2, we give a motivational example. Several models and basic concepts are introduced in Section 6.3. In Section 6.4, we propose our $LSHAP$ algorithm. We give the related work in Section 6.5. The experimental results are shown in Section 6.6, and the conclusion is shown in Section 6.7.

6.2 Motivational Examples

For the data mining engine, the working procedures are as follows. We first build up a execution time history table (ETHT). Then implement data cleaning. Next, do data integration, transformation, and filtering (selection). Finally we find the association rule of the random variables and the pattern of distribution function of each execution time.

Data mining engine works at compile time. Based on the obtained distribution function of each execution time, we will use loop scheduling to find the best assignment for cost minimization. Then at run time, needed data will be prefetched in advance. Based on the computed best assignment, we can prefetch data in certain time ahead with guaranteed probability. For example, if node A select type $F1$, then we prefetch data for node B in 2 time unit in advance, and it will guarantee with 100% that node B can be executed on time with needed data.

(a)

Nodes	F1 T	F1 P	F1 C	F2 T	F2 P	F2 C	F3 T	F3 P	F3 C
1	1	0.8	9	3	0.9	3	5	0.7	1
	2	0.2		4	0.1		6	0.3	
2	1	1.0	8	2	1.0	6	3	1.0	2
3	1	1.0	8	2	1.0	6	3	1.0	2
4	1	0.7	8	2	0.9	4	5	0.9	2
	3	0.3		4	0.1		6	0.1	
5	1	0.9	10	3	0.8	3	5	0.8	1
	2	0.1		4	0.2		6	0.2	

(b)

Nodes	F1 T	F1 P	F1 C	F2 T	F2 P	F2 C	F3 T	F3 P	F3 C
1	1	0.8	9	3	0.9	3	5	0.7	1
	2	1.0		4	1.0		6	1.0	
2	1	1.0	8	2	1.0	6	3	1.0	2
3	1	1.0	8	2	1.0	6	3	1.0	2
4	1	0.7	8	2	0.9	4	5	0.9	2
	3	1.0		4	1.0		6	1.0	
5	1	0.9	10	3	0.8	3	5	0.8	1
	2	1.0		4	1.0		6	1.0	

Figure 6.2. (a) A PDFG. (b) Parameter table

Assume an input PDFG (*Probability Data Flow Graph*) shown in Figure 6.2(a). Each node can select one of the three different FUs: F_1, F_2, and F_3. The execution times (T), and expected cost (C) of each node under different FU types are shown in Figure 6.2(b). The input PDFG has five nodes. Node 1 is a multi-child node, which has two children: 2 and 3. Node 5 is a multi-parent node, and has two parents: 3 and 4. The execution time T of each node is modeled as a random variable. For example, When choosing R_1, node 1 will be finished in 1 time unit with probability P_{111} and will be finished in 2 time units with probability P_{112}. The corresponding probabilities (P) to T are still unknown. P_{111} represents the first kind of variation of execution time for node 1 under type $F1$.

Data mining engine is used to predict the probabilities. We first profile the data and build the historic table of execution time of nodes. Then we use data mining techniques to discover the distribution function of each execution time that is not fix. Figure 6.3 (a) shows the obtained distribution of each random variable. Base on (a), Figure 6.3 (b) computes the cumulative distributed function (CDF) of each random variable.

For initial schedule graph in Figure 6.4 (c), we use the *HAP_Heu* algorithm in [70] to generate the minimum total cost with computed confidence probabilities under the timing

(a) Parameter table after data mining

Nodes	F1 T	F1 P	F1 C	F2 T	F2 P	F2 C	F3 T	F3 P	F3 C
1	1	0.8	9	3	0.9	3	5	0.7	1
	2	0.2		4	0.1		6	0.3	
2	1	1.0	8	2	1.0	6	3	1.0	2
3	1	1.0	8	2	1.0	6	3	1.0	2
4	1	0.7	8	2	0.9	4	5	0.9	2
	3	0.3		4	0.1		6	0.1	
5	1	0.9	10	3	0.8	3	5	0.8	1
	2	0.1		4	0.2		6	0.2	

(b) Parameter table with cumulative probabilities

Nodes	F1 T	F1 P	F1 C	F2 T	F2 P	F2 C	F3 T	F3 P	F3 C
1	1	0.8	9	3	0.9	3	5	0.7	1
	2	1.0		4	1.0		6	1.0	
2	1	1.0	8	2	1.0	6	3	1.0	2
3	1	1.0	8	2	1.0	6	3	1.0	2
4	1	0.7	8	2	0.9	4	5	0.9	2
	3	1.0		4	1.0		6	1.0	
5	1	0.9	10	3	0.8	3	5	0.8	1
	2	1.0		4	1.0		6	1.0	

Figure 6.3. (a) Parameter table after data mining. (b) Parameter table with cumulative probabilities.

constraints. The results are shown in Table 6.1. Algorithm *HAP_Heu* [70] is used as a sub-algorithm of our *LSHAP* algorithm. The entries with probability that is equal to 1 (see the entries in boldface) actually give the results to the hard real-time problem which shows the worst-case scenario. For each row of the table, the C in each (P, C) pair gives the minimum total cost with confidence probability P under timing constraint j. For example, using our algorithm, at timing constraint 11, we can get (0.90, 20) pair. The assignments are shown as "Ass_1" in Table 6.2. Assignment $A(u)$ represents the type selection of each node v. Hence, we find the way to achieve minimum total cost 20 with probability 0.90 satisfying timing constraint 11. While using the ILP and heuristic algorithm in [80], the total cost obtained is 32. The assignments are shown as "Ass_2" in Table 6.2.

For the new schedule graph shown in Figure 6.5(c), we get (0.90, 10) and (1.00, 12) pairs at timing constraint 11. For (0.90, 10) pair, node 5's type was changed to F_3, then the T was changed from 4 to 6, and cost change from 3 to 1. Hence the total execution time is 11, and the total cost is 10. So the improvement of cost saving is 50.0% while the probability is still 90%. For (1.00. 12) pair, the execution times of nodes 2, 3 were changed from 1 to be 3 and node 1's was changed from 2 to 6. The total cost were changed to 12, and the total execution time is still 11. Hence compared with original 32, the improvement

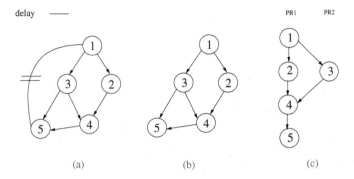

Figure 6.4. (a) Original PDFG. (b) The static schedule. (c) The schedule graph

of total cost saving is 62.5%. If we consider the cost of switch activity, we can get more practical results. For example, after rotation once, node 1 has changed from processor $PR1$ to $PR2$. Assume the cost of this switch is 1, then the final total cost is 13. Then the cost saving is 59.4% compared with previous scheduling and assignment.

6.3 Models and Concepts

In this section, we introduce the data mining and prefetching model, the system model, and the HAP problem.

Data Mining and Prefetching Model: Our data mining engine has several major steps [30], which are shown in Figure 6.6: First, build up an execution time history table (ETHT). In each iteration, the engine will store the execution time of each node into the table. From execution time history table, do data cleaning to remove noise and inconsistent data. Then do data integration and transformation. Next use data mining to extract data pattern. Finally, we obtain the probability distribution function of each execution time

Table 6.1. Minimum C satisfying T with P.

T	(P , C)	(P , C)	(P , C)	(P , C)	(P , C)
4	0.50, 43				
5	0.65, 39				
6	0.65, 35	0.81, 39			
7	0.65, 27	0.73, 33	0.81, 35	0.90, 39	
8	0.81, 27	0.90, 35	**1.00, 43**		
9	0.58, 20	0.73, 21	0.81, 27	0.90, 32	**1.00, 39**
10	0.72, 20	0.81, 21	0.90, 28	**1.00, 36**	
11	0.65, 14	0.90, 20	**1.00, 32**		
12	0.81, 14	0.90, 20	**1.00, 28**		
13	0.65, 12	0.90, 14	**1.00, 20**		
14	0.81, 12	0.90, 14	**1.00, 20**		
15	0.50, 10	0.90, 12	**1.00, 14**		
16	0.72, 10	0.90, 12	**1.00, 14**		
17	0.90, 10	**1.00, 12**			
18	0.50, 8	0.90, 10	**1.00, 12**		
19	0.72, 8	**1.00, 10**			
20	0.90, 8	**1.00, 10**			
21	**1.00, 8**				

and the association rules between inputs and execution time selection. For example, in the Figure 6.3, we obtained $P_{111} = 0.8$ and $P_{112} = 0.2$, which means node 1 with FU type 1 will be finished in 1 time unit with 80% probability and be finished in 2 time unit with 20% probability. We can also obtain the hidden association between different inputs and execution times. For instance, we may find node 1 will always be finished in 2 time unit when the input is a large integer m, such as $m \geq 100$. We may not know why this happened, but we can use this hidden association to select the right execution time and FU type.

For prefetching, suppose there is an on-chip memory and a main memory. The on-chip memory is small but fast; and the main memory is large but slow. Since the limited space of on chip memory, the engine will prefetch data that will be needed for later tasks based on the prediction of execution time of current task. For example, if the prefetching operation time is 2 cycles, and the current task node A is estimated to be finished in 5 cycles

Table 6.2. The assignments with $T = 11$.

		Node id	T	F	P	C
Ass_1	$A(v)$	1	2	F_1	1.00	9
		2	3	F_3	1.00	2
		3	3	F_3	1.00	2
		4	2	F_2	0.90	4
		5	4	F_2	1.00	3
	Total		11		0.90	20
Ass_2	$A(v)$	1	2	F_1	1.00	9
		2	1	F_1	1.00	8
		3	1	F_1	1.00	8
		4	4	F_2	1.00	4
		5	4	F_2	1.00	3
	Total		11		1.00	32

from now, then the engine need to prefetch 3 cycles before now the data for the next task, node B.

System Model: *Probabilistic Data-Flow Graph* (PDFG) is used to model applications of embedded systems. A cyclic *PDFG G* $=\langle$U, ED, d, T, F$\rangle$ is a node-weighted and edge-weighted directed graph, where $U = \langle u_1, \cdots, u_i, \cdots, u_N \rangle$ is the set of nodes; $F = \langle F_1, \cdots, F_j, \cdots, F_M \rangle$ is a FU set; the execution time $T_{F_j}(u)$ is a random variable; *ED* $\subseteq$ U $\times$ U is the edge set that defines the precedence relations among nodes in U. d(ed) is a function to represent the number of delays for any edge ed $\in$ ED, The edge without delay represents the intra-iteration data dependency; the edge with delays represents the inter-iteration data dependency and the number of delays represents the number of iterations involved. *Static Schedules:* From the PDFG of an application, we can obtain a static schedule. A static schedule of a cyclic PDFG is a repeated pattern of an execution of the corresponding loop. The DAG *directed acyclic graph* is obtained by removing all edges with delays in the original cyclic PDFG.

An *assignment A* is a function from domain U to range F, where U is the node set and F is the FU set. For a node $u \in V$, $A(u)$ gives selected FU type of node u. In a PDFG

delay ———

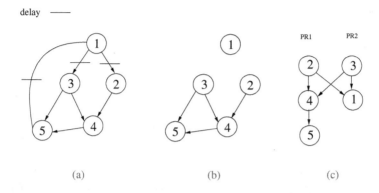

(a)   (b)   (c)

Figure 6.5. (a) The PDFG after retiming (b) The static schedule. (c) New schedule.

Figure 6.6. The basic structure of our method

G, $T_{F_j}(u)$, $1 \le j \le M$, represents the execution times of each node $u \in V$ when running at FU type F_j; $T_{F_j}(u)$ is either a discrete random variable or a continuous random variable. We define $X(t)$ to be the *cumulative distribution function* (abbreviated as *CDF*) of the random variable $T_{F_j}(u)$, where $X(t) = P(T_{F_j}(v) \le t)$. When $T_{F_j}(u)$ is a discrete random variable, the CDF $X(t)$ is the sum of all the probabilities associating with the execution times that are less than or equal to t. If $T_{F_j}(u)$ is a continuous random variable, then it has a *probability density function (PDF)*. If assume the PDF is f, then $X(t) = \int_0^t f(s)ds$. Function $X(t)$ is nondecreasing, and $X(-\infty) = 0$, $X(\infty) = 1$.

Retiming [49] is an optimal scheduling technique for cyclic DFGs considering inter-iteration dependencies. It can be used to optimize the cycle period of a cyclic PDFG by evenly distributing the delays. Retiming generates the optimal schedule for a cyclic PDFG when there is no resource constraint. Given a cyclic PDFG G=⟨U, ED, d, T, F⟩, retiming r of G is a function from U to integers. For a node $u \in U$, the value of r(u) is the number of delays drawn from each of incoming edges of node u and pushed to all of the outgoing edges. Let $G_r = \langle$U, ED, d_r, T, F$\rangle$ denote the retimed graph of G with retiming r, then $d_r(ed) = d(ed) + r(u) - r(v)$ for every edge $ed(u \rightarrow v) \in$ ED. *Rotation Scheduling* [15] is a scheduling technique used to optimize a loop schedule with resource constraints. It transforms a schedule to a more compact one iteratively in a PDFG. In most cases, the minimal schedule length can be obtained in polynomial time by rotation scheduling.

Definitions: Define the *HAP (heterogeneous assignment with probability)* problem as follows: Given M different FU levels: $F_1, F_2, \cdots, F_M$, a PDFG $G = \langle$ U, ED, d, T, F $\rangle$ with $T_{F_j}(u)$, $P_{F_j}(u)$, and $C_{F_j}(u)$ for each node $u \in$ U executed on each FU type F_j, a timing constraint L and a confidence probability P, find the FU for each node in assignment A that gives the *minimum expected total cost C with confidence probability P under timing constraint L.*

6.4   The Algorithms

To solve the HAP problem, we propose a high efficient algorithm, *LSHAP*, to minimize the total cost while satisfying timing constraints with guaranteed probabilities. The LSHAP algorithm is shown in Algorithm 6.4.1. In *LSHAP* algorithm. Algorithm *HAP_Heu* [70] is used as a sub-algorithm of our *LSHAP* algorithm.

Our *LSHAP* algorithm includes three major steps. First, we build up the *execution time history table* (ETHT) and use data mining to predict the execution time pattern from

---

Algorithm 6.4.1 *LSHAP* Algorithm

Input: M different FU levels, a PDFG, the timing constraint L, and the rotation times Q.

Output: An optimal voltage assignment to minimize cost while satisfying L

1: Build up the Execution Time History Table (ETHT).
2: Use data mining to predict the PDF of each varied execution time.
3: Build up the probability distribution function table and CDF table for each varied execution time.
4: Find the association rules between execution times and different inputs.
5: rotation $\leftarrow$ 0;
6: while (rotation $< Q$)
7:     Get the static schedule (a DAG) from input PDFG.
8:     Get scheduling graph from the DAG.
9:     Using algorithm *HAP_Heu* [70] to get the near optimal assignment of FU types for the schedule graph.
10:    Using Rotation scheduling [15] to retime the original PDFG and rotate down the first row.
11:    rotation++;
12: Output results: retime r, assignment A, and the minimum total cost E_{min}.
13: Use online prefetching to reduce cost while satisfying timing constraints with guaranteed probability.

ETHT. Second, we use an high efficient method to obtain the assignment with minimized total cost. This includes finding the FU assignment with cost minimization in each iteration and using rotation scheduling to obtain the best one for Q iterations. Finally, we use prefetching to prepare data in advance at run time. We compare our *LSHAP* algorithm with list scheduling and the ILP techniques in [110] on a set of benchmarks. The experimental data show that our *LSHAP* algorithm can get better results on cost reduction than that of the previous work.

In finding the FU assignment with cost minimization in each iteration, we use dynamic programming method traveling the graph in a bottom up fashion. For the ease of explanation, we will index the nodes based on bottom up sequence. Given the timing constraint L, a PDFG G, and an assignment A, we first give several definitions as follows:

1. G^i: The sub-graph rooted at node u_i, containing all the nodes reached by node u_i. In our algorithm, each step will add one node which

becomes the root of its sub-graph.

2. $C_A(G^i)$ and $T_A(G^i)$: The total cost and total execution time of G^i under the assignment A. In our algorithm, each step will achieve the minimum total cost of G^i with computed confidence probabilities under various timing constraints.

3. In our algorithm, table $D_{i,j}$ will be built. Here, i represents a node number, and j represents time. Each entry of table $D_{i,j}$ will store a link list of (Probability, Cost) pairs sorted by probability in an ascending order. Here we define the *(Probability, Cost) pair ($P_{i,j}$, $C_{i,j}$)* as follows: $C_{i,j}$ is the minimum cost of $C_A(G^i)$ computed by all assignments A satisfying $T_A(G^i) \le j$ with probability $\ge P_{i,j}$.

Usually, there are redundant pairs in a link list. We use Lemma 6.4.1 to cancel redundant pairs.

Lemma 6.4.1. *Given $(P_{i,j}^1, C_{i,j}^1)$ and $(P_{i,j}^2, C_{i,j}^2)$ in the same list:*

1. *If $P_{i,j}^1 = P_{i,j}^2$, then the pair with minimum $C_{i,j}$ is selected to be kept.*

2. *If $P_{i,j}^1 < P_{i,j}^2$ and $C_{i,j}^1 \ge C_{i,j}^2$, then $C_{i,j}^2$ is selected to be kept.*

In every step of our algorithm, one more node will be included for consideration. The information of this node is stored in local table $B_{i,j}$, which is similar to table $D_{i,j}$, but with cumulative probabilities only on node u_i. A local table store only data of probabilities and consumptions, of a node itself. Table $E_{i,j}$ is the local table storing only the data of node u_i. The building procedures of $B_{i,j}$ are as follows. First, sort the execution time variations in an ascending order for each R. Then, compute the CDF (cumulative distributive function) under each R. Finally, let $L_{i,j}$ be the link list in each entry of $B_{i,j}$, insert $L_{i,j}$ into $L_{i,j+1}$ while redundant pairs canceled out based on Lemma 6.4.1. We use the algorithm

HAP_Heu proposed by Qiu et al. [70] to solve FU assignment with cost minimization in each iteration, and algorithm *HAP_Heu* is used as a sub-algorithm of our *LSHAP* algorithm.

6.5   Related Work

Data Mining: Data mining is an analytic process designed to explore data in search of consistent patterns and/or systematic relationships between variables, and then to validate the findings by applying the detected patterns to new subsets of data. The ultimate goal of data mining is prediction. The process of data mining consists of three stages: (1) the initial exploration, (2) model building or pattern identification with validation/verification, and (3) deployment, i.e., the application of the model to new data in order to generate predictions.

Data mining is more oriented towards applications than the basic nature of the underlying phenomena [30]. In other words, Data mining is relatively less concerned with identifying the specific relations between the involved variables. For example, uncovering the nature of the underlying functions or the specific types of interactive, multivariate dependencies between variables are not the main goal of Data Mining. Instead, the focus is on producing a solution that can generate useful predictions.

Predictive data mining is the most common type of data mining and one that has the most direct applications. The goal of predictive data mining is to identify a statistical or neural network model or set of models that can be used to predict some response of interest [31, 100]. Text mining is another kind of popular data mining. While data mining is typically concerned with the detection of patterns in numeric data, very often important information is stored in the form of text. Unlike numeric data, text is often amorphous, and difficult to deal with. Text mining generally consists of two parts. One part is the analysis of multiple text documents by extracting key phrases, concepts, etc. The other part is the preparation of the text processed in that manner for further analyses with numeric data mining techniques, e.g., to determine co-occurrences of concepts, key phrases, names, addresses, product names, etc.

Prefetching: There are two general ways of instruction prefectching [34, 87, 93, 109]. The first one is called *Sequential instruction prefetching* . The simplest form of instruction prefetching is next line prefetching. In this scheme, when a cache line is fetched, a prefetch for the next sequential line is also initiated. A number of variations of this basic scheme exist, with different heuristics governing the conditions under which a prefetch is issued. Common schemes include: always issuing a prefetch for the next line (next-line always), issuing a prefetch only if the current line resulted in a miss (next-line on miss) and issuing a prefetch if the current line is a miss or is a previously prefetched line (next-line tagged). Next-N-line prefetch schemes extend this basic concept by prefetching the next N sequential lines following the one currently being fetched by the processor. The benefits of prefetching the next N-lines include, increasing the timeliness of the prefetches and the ability to cover short nonsequential transfers.

The other one is called *Nonsequential instruction prefetching* [44, 61, 85, 104]. Nonsequential prefetch prediction is closely related to branch prediction. history-based schemes, such as the target prefetcher, is one of the main styles of prefetcher specifically targeted at nonsequential prefetching, History table is a table is used to retain information about the sequence of cache lines previously fetched by the processor. As execution continues, the table is searched using the address of each demand fetched line. If the address hits in the table, the table returns the addresses of the next cache lines that were fetched the previous times the active cache line was fetched. Prefetches can then be selectively issued for these lines. In this scheme, the behavior of the application is predicted to be repetitious and prior behavior is used to guide the prefetching for the future requirements of the application. The prediction table may contain data for all transitions, or just the subset that relate to transitions between non-sequential cache lines.

6.6 Experiments

This section presents the experimental results of our algorithms. We build up a data mining engine and predict the probability distribution function of execution time at compile time. Then compute the best assignment and prefetch the data needed in advance. We compare our algorithm with list scheduling and the ILP techniques in [110] that can give near-optimal solution for the DAG optimization. Experiments are conducted on a set of benchmarks including 4-stage lattice filter, 8-stage lattice filter, voltera filter, differential equation solver, RLS-languerre lattice filter, and elliptic filter. M different FUs, $F_1, \cdots, F_M$, are used in the system, in which a processor under F_1 is the quickest with the highest cost and a processor under F_M is the slowest with the lowest cost. The execution times for each node are in Gaussian distribution. Each application has timing constraints. This assumption is good enough to serve our purpose to compare the relative improvement among different algorithms. For each benchmark, we conduct the experiments based on 5 processor cores. The experiments are performed on a Dell PC with a P4 2.1 G processor and 512 MB memory running Red Hat Linux 9.0.

The experimental results with 5 processors are shown in Table 6.3. In the table, columns "List", "ILP" , "LSHAP" represent the results obtained by list scheduling, the ILP in [110], our *LSHAP* algorithm, respectively. Columns "Cost" represents the cost and columns "%I" represents the improvement of our algorithm over the ILP in [110] with different probability. The total average cost-reduction improvement of *LSHAP* is shown in the last row.

From the experimental results, we can see that our algorithms achieve significant cost saving compared with the ILP in [110]. On average, *LSHAP* shows a 30.8% reduction

Table 6.3. The cost comparison for the schedules generated by list scheduling, the ILP in [110], and the LSHAP algorithm.

TC	List	ILP	LSHAP					
			0.8		0.9		1.0	
	Cost	Cost	Cost	%I	Cost	%I	Cost	%I
differential equation solver, 5 processors								
40	1018	865	432	50.1	506	41.5	602	30.4
50	887	755	382	49.4	438	42.0	533	29.4
60	826	702	352	49.8	389	44.6	471	32.9
70	643	643	331	48.5	362	43.7	455	29.2
80	706	600	303	49.5	348	42.0	409	31.8
4-stage lattice filter, 5 processors								
100	1986	1687	862	48.9	973	42.4	1178	30.2
110	1548	1316	661	49.8	754	42.7	915	30.5
120	1304	1108	579	47.8	604	45.5	756	31.8
130	1171	996	515	48.3	583	41.4	685	31.1
140	982	835	423	49.3	478	42.8	572	31.5
voltera filter, 5 processors								
80	1991	1692	869	48.6	973	42.5	1182	30.2
90	1368	1163	578	50.3	652	43.9	807	30.6
100	1233	1048	541	48.4	587	44.0	711	32.2
110	1164	989	501	49.4	571	42.3	696	29.6
120	1036	882	432	51.0	512	42.0	615	30.3
Average Imp. over ILP			49.3	–	42.9		–	30.8

in hard real-time, and reductions of 42.9%, 49.3% with probability 0.9 and 0.8, respectively, for soft real-time DSP systems. The reason of such big improvement is because we use loop scheduling to shrink the schedule length and assign the best possible FU to minimize cost while satisfying time constraint with guaranteed probabilities.

6.7 Conclusion

This chapter combined loop scheduling with data mining and prefetching to solve the heterogeneous FU assignment problem. We proposed an algorithm, *LSHAP*, to give the efficient solutions. We first used data mining to obtain the PDF of execution times, which were modeled as random variable. And found the association rules between execution times and different inputs. Then, by taking advantage of the uncertainties in execution time of tasks, we gave out FU assignments and scheduling to minimize the expected total cost while satisfying timing constraint with guaranteed probabilities. We repeatedly regrouped a loop based on rotation scheduling and decrease the energy by voltage selection as much as possible within a timing constraint. Finally, we prefetched the data needed in advance at run time. Our approach can handle loops efficiently for heterogeneous real-time embedded systems. Experimental results on a vast range of benchmarks showed the significant cost-saving of our approach.

CHAPTER 7

TYPE ASSIGNMENT AND SCHEDULING WITH VARIABLE PARTITION TO MINIMIZE ENERGY FOR HETEROGENEOUS MULTI-BANK MEMORY

Many high-performance DSP processors employ multi-bank on-chip memory to improve performance and energy consumption. This architectural feature supports higher memory bandwidth by allowing multiple data memory accesses to be executed in parallel. However, making effective use of multi-bank memory remains difficult, considering the combined effect of performance and energy requirement. This chapter studies how to minimize the total energy consumption while satisfying timing constraint on heterogeneous multi-bank memory. An algorithm, TAMRS (*Type Assignment and Minimum Resource Scheduling*), is proposed. The algorithm use type assignment with the consideration of variable partition to find the best configuration for both memory and ALU. The experimental results demonstrate the effectiveness on energy reduction of our TAMRS algorithm.

7.1 Introduction

Memory access time latency and energy consumption are two of the most important design considerations in memory architecture. A number of papers have investigated how to exploit multi-bank memory from either performance or energy aspect. But the combined effect of both performance and energy requirement is seldom tackled. In high-performance *digital signal processing* (DSP) applications, strict real-time processing is critical [112] since the growing speed gap between *central processing unit* (CPU) and memory becomes a bottleneck for designing such real-time systems. To close this speed gap, embedded systems need to utilize multi-bank on-chip memories [53, 98, 99]. The high energy consumption of memories make them target of many energy-conscious optimization techniques [12]. This is especially true for mobile applications, which are typically memory-intensive. This chapter focuses on reducing the total energy while satisfying performance constraints for loop applications on multi-bank memory architectures.

In many advanced memory architecture, there are heterogeneous memory banks. Different memory banks have different memory access time latency and energy consumption for same operations [4, 26, 42, 80]. A certain memory type may access the data stored slower but with less energy consumption, while another type will access the data faster with higher energy consumption. Also, there is a limitation of how many banks can be accessed simultaneously in certain memory architectures. Therefore, an important problem arises: how to assign types to the banks selected and partition variables for an application to minimize the total energy consumption while satisfying timing constraints.

Much research has been conducted in the area of using multi-bank memory to achieve maximum instruction level parallelism, i.e., optimize performance [21, 51, 54, 78, 88, 103]. These approaches differ in either the models or the heuristics. However, these works seldom consider the combined effect of performance and energy requirements. Actually, performance requirement often conflicts with energy saving [22, 23, 48, 63, 94, 106]. Hence, significant energy saving and performance improvements can be obtained by exploiting heterogeneous multi-bank memory at the instruction level. Wang et al. [97] have considered the combined effect and proposed the VPIS algorithm to overcome it, but their algorithm does not fully exploit the heterogeneous multi-bank memory architecture.

Combining the consideration of energy and performance for both memory bank and ALU, in the chapter, we propose a novel graph model to overcome the weakness of previous techniques. We design an algorithm, TAMRS (*Type Assignment and Minimum Resource*

Scheduling), to minimize total energy while satisfying performance requirements. The experimental results show that TAMRS achieves a significant reduction on average in total energy consumption. For example, with 3 memory types and 3 ALU types, compared with the VPIS algorithm in Wang et al.'s paper [97], TAMRS shows an average 15.6% reduction in total energy consumption.

In summary, this chapter has several major contributions: First, we study the combined effects of energy-saving and performance of memory and ALU in a systematic approach. Second, we exploit the energy-saving with type assignment and minimum resource scheduling for both memory and ALU. Third, data locality has been improved by using variable partition.

In the next section, we introduce basic concepts and models. An example is given in Section 7.3. The algorithm is discussed in Section 7.4. We show our experimental results in Section 7.5. Related work and concluding remarks are provided in Section 7.6 and 7.7, respectively.

7.2 Basic Concepts and Models

In this section, we introduce some basic concepts which will be used in the later sections. First, the *data flow graph* (DFG) is given to model heterogeneous multi-bank memory and multi-type ALU architecture. Next, we introduce the concepts of variable partition and *Variable Independence Graph* (VIG). Finally, we give the formal definition of heterogeneous multi-bank type assignment problem.

7.2.1 Data Flow Graph

Data Flow Graph (DFG) is used to model many multimedia and DSP applications. The definition is as follows:

Definition 7.2.1. *A DFG G = ⟨U, ED, T, E⟩ is a node-weighted directed acyclic graph (DAG), where $U = \langle u_1, \cdots, u_i, \cdots, u_N \rangle$ is a set of operation nodes; $ED \subseteq U \times U$ is an edge set that defines the precedence relations among nodes in U; T is a set of operation time for all nodes in U; E is a set of energy consumption for all nodes in U.*

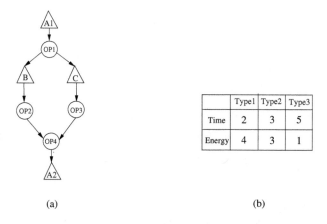

	Type1	Type2	Type3
Time	2	3	5
Energy	4	3	1

(a)   (b)

Figure 7.1. (a) A DFG with both memory and ALU operations. (b) The types of memory.

The nodes in the data flow graph can be either memory operations or ALU operations. As shown in Figure 7.1 (a), the ALU operations are represented by circles, and the memory operations are represented by triangles. Particularly, an edge from a memory operation node to an ALU operation node represents a *Load* operation, whereas the edge from an ALU operation node to a memory operation node represents a *Store* operation. The memory node with same alphabet prefix but different number postfix stands for the same variable but different memory access operations. For instance, node $A1$ and $A2$ are memory operations accessed by the same variable A, but they are different memory operations.

Figure 7.1 (a) shows a DFG. We start from loading value of variable A into ALU to do operation 1, i.e., $OP1$. Then load variable B to do $OP2$. The same is for $OP3$. Next, do $OP4$ with the inputs of the results of $OP2$ and $OP3$. Finally, store the result of $OP4$ into variable A in memory. In this example, we have 2 same type ALUs, they can finish each operation in 1 clock cycle with energy consumption 1 μJ. Assume there are 3 types of banks to be chosen from and we can only use maximum 2 types of banks. Only 2 banks can be accessed at the same time. The memory types are shown in Figure 7.1 (b). Type 1 has memory access time latency 2 clock cycles with energy consumption 4 μJ; Type 2 has memory access time latency 3 with energy consumption 3; And the time is 5 and energy is 1 for type 3. We can represent them in Type(Time, Energy) format, such as 1(2, 4). There is a timing constraint L and it must be satisfied for executing the whole DFG, including both memory access part and ALU part.

7.2.2   Variable Partition and VIG

Variable partition [97, 112] is an important method to improve the data locality. Different variable partitions will significantly affect the schedule length and the energy consumption of an application. In order to properly partition variables, we use VIG (*Variable Independence Graph*) to expose all parallel memory accesses in a DFG [112]. The nodes of the graph represent variables, and the edges in the graph represent potential parallelism existing among the memory accesses for these variables.

Definition 7.2.2. *A VIG is an undirected weighted graph $G_v = \langle U, ED, w \rangle$, where U is a set of nodes representing variables, and $ED \subseteq U \times U$ is a set of edges connecting between nodes in U, whose memory operations can be executed in parallel potentially. Function w maps from ED to a set of real values representing a priority of partitioning nodes u and v to different memory banks of an edge $u \rightarrow v \in ED$.*

To capture the tradeoff between the desire of parallelism and that of serialism, Wang et al. [97] use two lists of weights. One is the list of possibility weights, which are referred to as parallelism weights. The other one is the list of weights that are referred to as serialism weights. The goal of serialism weights is to model the possibility of serializing a pair of operations without sacrificing performance. In this paper, we use both parallelism weight in [112] and serialism weight in [97] in building the VIG.

For example, assume there are two memory banks and two ALUs. Both banks are in type 1 (2, 4). Figure 7.2(a) shows the schedule 1 with B and C in different banks. Thus we can fully take advantage of parallelism by assigning variable B to M1 and variable C to M2 at the same time unit 4 and 5. The schedule length is only 9. When we group B and C into the same bank, then the schedule length is changed to be 11. We must use B and C in serial, and can not take advantage of the parallelism. The corresponding schedule 2 is shown in Figure 7.2(b). Therefore, variable partition will affect the schedule length and the total time and energy consumption for the DFG of an application.

Time	ALU1	ALU2	M1	M2
1			A1	
2			A1	
3	OP1			
4			B	C
5			B	C
6	OP2	OP3		
7	OP4			
8			A2	
9			A2	

(a)

Time	ALU1	ALU2	M1	M2
1			A1	
2			A1	
3	OP1			
4			B	
5			B	
6	OP2		C	
7			C	
8	OP3			
9	OP4			
10			A2	
11			A2	

(b)

Figure 7.2. (a) Schedule 1 with B and C in different banks. (b) Schedule 2 with B and C in the same bank.

In the following, we introduce some important concepts that will be used in constructing a complete VIG. During the graph construction, we will be particularly interested in some memory operation pairs that help us identify the parallel memory accesses. We call them "independent pairs". For example, the nodes B and C in Figure 7.1(a) are independent pairs.

Definition 7.2.3. *Given a DFG $G = \langle U, ED, T, E \rangle$, if nodes $u, v \in V$, are not reachable from each other through any path without delay in the DFG G, then nodes u and v are independent pairs.*

The mobility property of a node in a schedule is very important in improving the preciseness in the graph construction. Given a DFG $G = \langle U, ED, T, E \rangle$, a *mobility window* [58] of node $v \in U$, which is denoted by $MW(v)$ in this chapter, is a set of time units in a static schedule by which node v can be placed. The first time unit where the node v can be scheduled is determined by as soon as possible (ASAP) scheduling, and the last control step by which node v can be scheduled is determined by as late as possible (ALAP) scheduling with the longest path as a time constraint. Mobility window gives the earliest and the latest position a node can be scheduled. Note that the overlap of mobility windows of two nodes indicates the possibility that the nodes could be scheduled in the same time unit.

In the following, we define the priority function of an edge in VIG based on mobility windows of two parallel memory accesses. We use the cardinality of mobility window overlap to denote the possible occurrences of parallel operations and use the multiplication of the cardinalities of two mobility windows to denote all arrangements of two nodes in a schedule. We define the variable partition problem as follows:

Definition 7.2.4. *Given a VIG $G_v = \langle U, ED, w \rangle$, and let n to be the number of partitions required, the variable partitioning problem is to partition U into n disjoint sets $P_1, P_2, \cdots, P_2$, such that the total $w(u, v)$, $\forall u \in P_i, v \in P_j$, $\forall i, j = 1, \cdots, n$, is maximum.*

A VIG can be built in various ways, depending on how accurately the graph conveys the potential memory access parallelism in the program. Different graph constructions can lead to different variable partitioning results. For the variable partitioning problem that is aimed to produce shorter schedule, the accuracy of the VIG is limited by the unknown positions of the memory operations in a schedule. We would like to provide a complete and accurate view for variable partitioning as possible, but on the other hand, we also would like to keep the flexibility so that the partitioning process can work with different scheduling algorithms. The intricacy of building the graph model for the variable partitioning problem is how to keep certain level of accuracy of the parallelism and still have a graph working for the variable partitioning problem in an effective way. We first give two intuitive ways to build the initial VIG graph.

Construction of VIG-1: Given a DFG $G = \langle U, ED, T, E \rangle$, if there exists a pair of memory operation nodes u and v that are independent pairs in G, then there is an edge (u, v) in the VIG.

Construction VIG-2: Given DFG $G = \langle U, ED, T, E \rangle$, if there exists a pair of memory nodes u and v that are independent pairs in G, and $MW(u) \cap MW(v) \neq \emptyset$, then there is an edge (u, v) in the VIG.

The construction of the VIG graph is based on the DFG representation of an application. From the DFG representation of a program, one can readily derive both the as soon as possible (ASAP) and as late as possible (ALAP) schedules, considering the constraints of computation units. Let the control steps of a memory operation, a, be $t_s(a)$ and $t_l(a)$ according to ASAP and ALAP, respectively. The mobility, that is, the scheduling freedom of a, defined as $[t_s, t_l]$, represents the time interval in which a can be scheduled without introducing additional delay. Only when the mobilities of two memory operations have some overlap may parallelizing the two corresponding variables be beneficial, in terms of

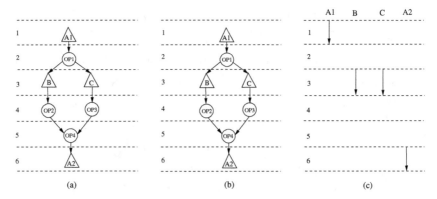

Figure 7.3. (a) ASAP schedule. (b) ALAP schedule. (c) Memory operation mobility graph.

improving performance. Clearly, the larger the overlap between two mobilities, the higher the potential of the two variables being able to be parallelized. If the mobilities of two operations are both small and their overlap is relatively large, parallelizing the corresponding variables is more likely to improve the schedule length. In other words, if such variables are put in the same bank, accessing the two variables is forced to be sequentialized, which is very likely to increase the overall schedule length. Zhuge et al. [112] assigned a possibility weight defined below to an edge to model this property.

Definition 7.2.5. *Given two memory operations, a and b, let their mobilities be $[t_s(a), t_l(a)]$ and $[t_s(b), t_l(b)]$, and the maximum overlap between these two mobilities be the interval $[t_1, t_2]$. The possibility weight assigned to the edge between the two variables accessed in operations a and b is $\frac{t_2 - t_1 + 1}{(t_l(a) - t_s(a) + 1)(t_l(b) - t_s(b) + 1)}$.*

For example, in Figure 7.3(a), for the DFG in Figure 7.1(a), we compute the mobility of each variable by using the ASAP and ALAP schedules of the DFG. Then we show the ALSP schedule in Figure 7.3(b). We compute the mobility of each variable. The we draw the mobility graph in Figure 7.3(c). For instance, Variable B and C has $MW(B) = [3, 3]$ and $MW(C) = [3, 3]$. B and C are independent pairs in VIG, using *construction of VIG-1*. Base on *construction of VIG-2* and Definition 7.2.5, the pair (B,C) has weight 1; and there is an edge (B, C) in VIG graph. There is no edge between the two nodes of pairs (A1, B), (A1, C), (A2, B), and (A2, C).

7.2.3 Heterogeneous Bank Energy Saving Problem

An *assignment* A is a function from domain U to range R, where U is the node set and R is the type set. For a node $u \in U$, $A(u)$ gives selected mode of node u. In a DFG G, $T_{R_j}(u)$, $1 \leq j \leq M$, represents the execution times of each node $u \in U$ when running with type R_j; For each type R_j with respect to node u, there is a set of E_i, which is the energy consumption of each node in DFG. $E_{R_j}(u)$, $1 \leq j \leq M$, is used to represent the energy consumption of each node $u \in U$ on mode R_j, $E_{R_j}(u) = \sum E_i$, which is a fixed value. Given an assignment A of a DFG G, we define the *system total energy consumption under assignment A*, denoted as $E_A(G)$, to be the summation of energy consumption, $E_{A(u)}(u)$, $u \in U$, of all nodes, that is, $E_A(G) = \sum_{u \in U} E_{A(u)}(u)$. In this chapter, we call $E_A(G)$ *total energy consumption* in brief.

Define the (*Heterogeneous Bank Energy Saving*) problem as follows: Given M different types: $R_1, R_2, \cdots, R_M$, a DFG $G = \langle U, ED, T, E \rangle$ with $T_{R_j}(u)$ and $E_{R_j}(u)$ for each node $u \in U$ executed on each type R_j, a timing constraint L, find a type assignment A using only K types to give the minimum energy consumption E while satisfying timing constraint L.

7.3 Motivational Example

In this section, we continue the example in Figure 7.1(a) and give the final solution by using the algorithm we propose.

From variable partition, we know B and C should be in different banks. Also $A1$ and $A2$ should be in same bank since they have same prefix "A", i.e., different operations corresponding to same variable. Based on precedence relations in Figure 7.1 (a) and variable partition information, we do type assignment to minimize total energy under different timing constraints L. For instance, under timing constraint 10, the type assignment for memory operations is: $A1, B, A2$ in one bank with type 1(2,4) and C in another bank with type 2(3, 3). The total time for memory part is 7, and the total energy consumption for memory part is 15. In this example, the DFG of memory is shown in Figure 7.4 (a). Adding up the ALU part, which have total time 3 and total energy consumption 4, we get the fianl result: the total time is 10 and the total energy consumption is 19. The result is shown in Figure 7.4 (b).

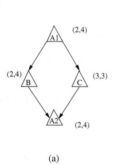

Time	ALU1	ALU2	M1	M2
1			A1	
2			A1	
3	OP1			
4			B	C
5			B	C
6		OP2		C
7		OP3		
8	OP4			
9				A2
10				A2

(a)   (b)

Figure 7.4. (a) The memory DFG. (b) The result of *Type_Assign* for the DFG in Figure 7.1(a) with timing constraint 10.

After the type assignment, we implement minimum resource scheduling. For example, at timing constraint 10, we only need one ALU and two memories. One memory is type 1, and the other is type 2. In Figure 7.4 (b), we can operations $OP2$ and $OP3$ to column ALU1 and let ALU1 to execute them.

Time	ALU1	ALU2	M1	M2
1			A1	
2			A1	
3	OP1			
4			B	C
5			B	C
6	OP2			C
7	OP3			
8	OP4			
9			A2	
10			A2	

Figure 7.5. The minumum resource scheduling for the DFG in Figure 7.1(a) with timing constraint 10.

Based on Figure 7.1(a), we obtained the minimum energy consumptions satisfying different timing constraints by using our algorithm. The results are shown in Table 7.1. "T" represents total time spent and "E" represents total energy consumption of the DFG. There are only total five solutions with different timing constraints based on the restrictions of different inputs.

Table 7.1. The results of *Type_Assign* for the DFG in Figure 7.1(a).

$Time$	9	10	12	14	18
(E_i)	20	19	16	14	8

7.4 The Algorithms

In this section, an algorithm, TAMRS (*Type Assignment and Minimum Resource Scheduling*), is designed to solve the problem of minimizing total energy without sacrificing performance.

7.4.1 The TAMRS Algorithm

Algorithm 7.4.1 TAMRS

Input: DFG $G = \langle U, ED, T, E \rangle$ with memory and ALU operations, $N1$ types of memory banks, $N1$ types of ALUs, each type has (energy E, latency T) attributes. K number of memory banks that can be accessed simultaneously, P numbers of the ALUs, the timing constraint L.

Output: type assignment A for each bank to minimize energy E while satisfying L.

1: Build VIG graph for variable partition, find both variable partition weights;
2: Use dynamic programming *Type_Assign* to get assignments with at most K types with the consideration of variable partition weight for memory part;
   And get assignments with at most P types for ALU part;
3: Do ALU and memory scheduling using *Minimum Resource Scheduling and Configuration* Algorithm;
4: $E_{min} \leftarrow$ the minimum total energy consumption;
5: Output E_{min} and corresponding A;

The TAMRS algorithm is shown in Algorithm 7.4.1. Based on Figure 7.1 (a), we obtained the variable partition weight by building VIG graph. Then we use $Type\_Assign$ algorithm to find the type assignments with at most K types of memory and P types of ALU while satisfying timing constraint L. Finally, we do memory and ALU scheduling using *Minimum Resource Scheduling and Configuration* Algorithm. The outputs include the minimum energy consumption, corresponding assignment, and the minimum resource scheduling.

TAMRS algorithm has combined several novel techniques to explore the heterogeneous type memory bank and ALU: First, we propose a novel *Type_Assign* algorithm, which use dynamic programming with the consideration of variable partition weights. Second, VIG graph has been built to obtain variable partition weights. Third, we consider heterogeneous type assignment and minimum resource scheduling and configuration for both memory and ALU.

7.4.2 Definitions and Lemma of Type_Assign

To solve the type assignment problem for a certain schedule S, we use dynamic programming method traveling the graph in a bottom up or top down fashion. For the ease of explanation, we will index the nodes based on bottom up sequence.

Given the timing constraint L, a DFG G, and an assignment A, we first give several definitions as follows:

1. The function from domain of variable to range of bank-type is defined as Bank(). For example, "Bank(A)= type 1" means the bank-type of variable A is type 1.

2. G^i: The sub-graph rooted at node u_i, containing all the nodes reached by node u_i. In our algorithm, each step will add one node which becomes the root of its sub-graph.

3. $E_A(G^i)$ and $T_A(G^i)$: The total energy consumption and total execution time of G^i under the assignment A. In our algorithm, each step will achieve the minimum total energy consumption of G^i under various timing constraints.

4. In our algorithm, table $D_{i,j}$ will be built. Here, i represents a node number, and j represents time. Each entry of table $D_{i,j}$ will store energy consumption $E_{i,j}$ and its corresponding linked list. Here we define $E_{i,j}$ as follows: $E_{i,j}$ is the minimum energy consumption of $E_A(G^i)$ computed by all assignments A satisfying $T_A(G^i) \leq j$. The linked list records the type selection of all previous nodes passed, from which we can trace back how $E_{i,j}$ is obtained.

We have the following lemma about energy consumption cancellation when the total execution time is same.

Lemma 7.4.1. *Given $E_{i,j}^1$ and $E_{i,j}^2$ with the same total execution time, if $E_{i,j}^1 \geq E_{i,j}^2$, then $E_{i,j}^2$ is selected to be kept.*

In each step of dynamic programming, we have several rules about cancellation of redundant energy consumption and its corresponding linked list.

- Rule 1: If the number of memory types greater than K or the number of ALU types greater than P, then discard the corresponding energy consumption and linked list.

- Rule 2: If two sibling a and b, i.e., children of same node, are not allowed to be same type, i.e., Bank(a) $\neq$ Bank(b), in variable partition, and Bank(a) = Bank(b) in assignment, then discard the corresponding E and linked list, except the scenario that all the nodes till now are in the same type, i.e., $\forall\, u,\, v \in G$, Bank(u) = Bank(v) .

- Rule 3: If two sibling a and b are just exchanged their types and other nodes are same in type assignment for the two corresponding linked list, then only keep one E and its corresponding linked list.

For example, in Figure 7.1(a), using bottom up approach, in the fourth step, i.e. for node A1, if we get a sequence (2,4)(3,3)(5,1) for A2, B, C nodes, then discard this sequence. In Figure 7.1(b), B and C are not allowed to be same type except all A, B, and C are in same type. If we get two sequence (2,4)(3,3)(3,3), then delete it. But we will keep the sequence (3,3)(3,3)(3,3). In Figure 7.1(c), at node A1, if we get two sequences (2,4)(3,3)(2,4) and (2,4)(2,4)(3,3), then base on symmetric rule of rule 3, we only keep one sequence.

In every step of our algorithm, one more node will be included for consideration. The data of this node is stored in local table $B_{i,j}$, which is similar to table $D_{i,j}$, but with energy consumption only on node u_i. A local table store only data of energy consumption

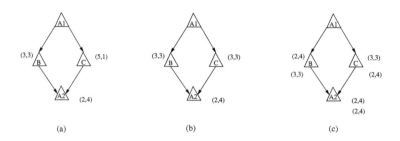

Figure 7.6. Three examples (a), (b), and (c).

of a node itself. Table $B_{i,j}$ is the local table only storing the energy consumption of node u_i. $E_{i,j}$ is the energy consumption only for node u_i with timing constraint j, The algorithm to compute $D_{i,j}$ are shown in Algorithm 7.4.2.

Algorithm 7.4.2 algorithm to compute $D_{i,j}$ for a simple path

---

Input: A simple path DFG
Output: $D_{i,j}$
1: Build a local table $B_{i,j}$ for each node;
2: Start from u_1, $D_{1,j} \leftarrow B_{i,j}$;
3: **for all** $u_i, i > 1$, **do**
4:     **for all** timing constraint j, **do**
5:         Compute the entry $D_{i,j}$ as follows:
6:         **for all** k in $B_{i,j}$, **do**
7:             $D_{i,j} = D_{i-1,j-k} + B_{i,k}$;
8:             Cancel redundant energy and linked list with the three redundant linked list cancellation rules;
9:             Insert $D_{i,j-1}$ to $D_{i,j}$ and remove redundant energy value according to Lemma 7.4.1;
10:         **end for**
11:     **end for**
12: **end for**

7.4.3 The Type_Assign Algorithm

In algorithm *Type_Assign*, which is shown in Algorithm 7.7, without loss of generality, assume using bottom up approach. Algorithm *Type_Assign* gives the near-optimal solution

Input: DFG $G = \langle U, ED, T, E \rangle$ with N nodes, M types of memory banks with (T, E) pairs. K number of memory banks that can be accessed simultaneously, P numbers of the ALUs, the timing constraint L.

Output: An efficient types assignment to MIN(E) while satisfying L.

Algorithm:

1. $SEQ \leftarrow$ Sequence obtained by topological sorting all the nodes.

2. $t_{mp} \leftarrow$ the number of multi-parent nodes; $t_{mc} \leftarrow$ the number of multi-child nodes;
   If $t_{mp} < t_{mc}$, **Then** use bottom up approach;
   Else, use top down approach.

3. **If** bottom up approach, **Then** use the following algorithm;
   If top down approach, **Then** just reverse the sequence.

4. $SEQ \leftarrow \{u_1 \rightarrow u_2 \rightarrow \cdots \rightarrow u_N\}$, in bottom up fashion;
   $D_{1,j} \leftarrow B_{1,j}$;
   $D'_{i,j} \leftarrow$ the table that stored MIN(E) for the sub-graph rooted on u_i except u_i;
   $u_{i_1}, u_{i_2}, \cdots, u_{i_W} \leftarrow$ all child nodes of node u_i; $w \leftarrow$ the number of child nodes of node u_i.

5. **If** w = 0, **Then** $D'_{i,j} = (0,0)$;
   If w = 1, **Then** $D'_{i,j} = D_{i_1,j}$;
   If w > 1, **Then** $D'_{i,j} = D_{i_1,j} + \cdots + D_{i_w,j}$;

6. Computing $D_{i_1,j} + D_{i_2,j}$:
   $G' \leftarrow$ the union of all nodes in the graphs rooted at nodes u_{i_1} and u_{i_2};
   Travel all the graphs rooted at nodes u_{i_1} and u_{i_2};
   If a node is a common node, **Then** use a selection function to choose the type of a node.
   Cancel redundant energy and linked list with the three redundant linked list cancellation rules.

7. For each k in $B_{i,k}$,
   $D_{i,j} = D'_{i,j-k} + B_{i,k}$

8. $D_{N,j} \leftarrow$ a table of MIN(E);
   Output $D_{N,L}$.

Figure 7.7. *Type_Assign* Algorithm.

when the given DFG is a DAG. In step 6, $D_{i_1,j} + D_{i_2,j}$ is computed as follows. let G' be the union of all nodes in the graphs rooted at nodes u_{i_1} and u_{i_2}. Travel all the graphs rooted at nodes u_{i_1} and u_{i_2}. If a node q in G' appears for the first time, we add the energy consumption of q to $D'_{i,j}$. If q appears more than once, that is, q is a common node, we only count it once. That is, the energy consumption is just added once. The final $D_{N,j}$ we get is the table in which each entry has the minimum energy consumption under the timing constraint j. In the following, we give the Theorem 7.4.1 about this.

Theorem 7.4.1. *For each $E_{i,j}$ in $D_{i,j}$ ($1 \leq i \leq N$) obtained by algorithm* Type_Assign, *$E_{i,j}$ is the minimum total cost for the graph G^i under timing constraint j.*

Proof. **By induction.** *Basic Step:* When $i = 1$, there is only one node and $D_{1,j} = B_{1,j}$. Thus, when $i = 1$, Theorem 7.4.1 is true. *Induction Step:* We need to show that for $i \geq 1$, if for each $E_{i,j}$ in $D_{i,j}$, $E_{i,j}$ is the minimum total energy consumption of the graph G^i, then for each $E_{i+1,j}$ in $D_{i+1,j}$, $E_{i+1,j}$ is the total energy consumption of the graph G^{i+1} under timing constraint j. According to the bottom up approach (for top down approach, just reverse the sequence), the execution of $D_{i,j}$ for each child node of v_{i+1} has been finished before executing $D_{i+1,j}$. From step 5, $D'_{i+1,j}$ gets the summation of the minimum total energy consumption of all child nodes of u_{i+1} because they can be executed simultaneously within time j. We avoid the repeat counting of the common nodes. Hence, each node in the graph rooted by node u_{i+1} was counted only once. From step 7, the minimum total energy consumption is selected from all possible energy consumption caused by adding u_{i+1} into the sub-graph rooted on u_{i+1}. So for each $E_{i+1,j}$ in $D_{i+1,j}$, $E_{i+1,j}$ is the total energy consumption of the graph G^{i+1} under timing constraint j. Therefore, Theorem 7.4.1 is true for any i ($1 \leq i \leq N$). $\qquad\square$

7.4.4 The Minimum Resource Scheduling and Configuration

We have obtained type assignment with at most K types of memory banks and P types of ALU. Then we will compute the configuration of the types and give the corresponding schedule which consume minimum energy while satisfying timing constraint. We propose minimum resource scheduling algorithms to generate a schedule and a configuration which satisfying our requirements. We first propose *Algorithm Lower_Bound_RC* that produces an initial configuration with low bound resource. Then we propose *Algorithm Min_RC_Scheduling* that refine the initial configuration and generate a schedule to satisfy the timing constraint.

Algorithm Lower_Bound_RC is shown in Figure 7.8. In the algorithm, it counts the total number of every type in every time unit in the ASAP and ALAP schedule. Then the lower bound for each bank type is obtained by the maximum value that is selected from the average resource needed in each time period. For example, for the DFG in Figure 7.1(a), after using *Min_RC_Scheduling* algorithm, we find that the lower bound of both ALU and memory are 1.

Using the lower bound of each type as an initial configuration, we propose an algorithm, *Algorithm Min_RC_Scheduling*, which is shown in Figure 7.9, to generate a schedule that satisfies the timing constraint and gets the finial configuration. In the algorithm, we first compute $ALAP(v)$ for each node v, where $ALAP(v)$ is the schedule step of v in the ALAP schedule. Then we use a revised list scheduling to perform scheduling. In each scheduling step, we first schedule all nodes that have reached to the deadline with additional resource if necessary and then schedule all other nodes as many as possible without increasing resource. For example, for the DFG in Figure 7.1(a), after using *Min_RC_Scheduling* algorithm, we find that only one ALU and two memory (one is type 1, the other is type 2) are needed. The scheduling is shown in Figure 7.1(b).

Input: A DFG with type assignments and timing constraint L.
Output: Lower bound for each type
Algorithm:

1. Schedule the DFG by ASAP and ALAP scheduling, respectively.

2. $N_{ASAP}[i][j] \leftarrow$ the total number of nodes with type j and scheduled in step i in the ASAP schedule.

3. $N_{ALAP}[i][j] \leftarrow$ the total number of nodes with type j and scheduled in step i in the ALAP schedule.

4. For each type j,
$$LB_{ASAP}[j] \leftarrow \max\{N_{ASAP}[1][j]/1,$$

$$(N_{ASAP}[1][j] + N_{ASAP}[2][j])/2,$$

$$\cdots, \sum_{1 \leq k \leq L} N_{ASAP}[k][j]/L\}$$

5. For each type j,
$$LB_{ALAP}[j] \leftarrow \max\{N_{ALAP}[L][j]/1,$$

$$(N_{ALAP}[L][j] + N_{ALAP}[L-1][j])/2,$$

$$\cdots, \sum_{1 \leq k \leq L} N_{ALAP}[L-k+1][j]/L\}$$

6. For each type j, its lower bound:
$$LB[j] \leftarrow \max\{LB_{ASAP}[j], LB_{ALAP}[j]\}$$

Figure 7.8. Algorithm Lower_Bound_RC.

Input: A DFG with type assignments, timing constraint L, and an initial configuration.
Output: A schedule and a configuration.
Algorithm:

1. For each node v in DFG, compute ALAP(v) that is the schedule step of v in the ALAP schedule.

2. $S \leftarrow 1$.

3. Do {

   - Ready_List $\leftarrow$ all ready nodes;
   - For each node $v \in$ Ready_List, if ALAP(v)==S, schedule v in step S with additional resource if necessary;
   - For each node $v \in$ Ready_list, schedule node v without increasing current resource or schedule length;
   - Update Ready_List and $S \leftarrow S + 1$;

   } **While ($S \leq L$);**

Figure 7.9. Algorithm Min_RC_Scheduling.

Algorithms *Lower_Bound_RC* and *Min_RC_Scheduling* both take $O(|U| + |ED|)$ to get results, where $|U|$ is the number of nodes and $|ED|$ is the number of edges for a given DFG.

7.5 Experiments

In this section, we conduct experiments with the TAMRS algorithm on a set of benchmarks including Wave Digital filter (WDF), Infinite Impulse filter (IIR), Differential Pulse-Code Modulation device (DPCM), Two dimensional filter (2D), Floyd-Steinberg algorithm (Floyd), and All-pole filter. The proposed run-time system has been implemented and a simulation framework to evaluate its effectiveness has been built. K different memory types, $M_1, \cdots, M_K$, are used in the system, in which a memory with type M_1 is the quickest with the highest energy consumption and a memory with type M_K is the slowest with the lowest energy consumption.

We conducted experiments using three different methods:

Method 1: uniform type + list scheduling;

Method 2: exhaustive type assignment + list scheduling;

Method 3: our TAMRS algorithm.

In Method 1, there is only one type memory, the type is the one with minimum energy consumption, i.e., type M_K. Based on homogeneous type, we do list scheduling. In Method 2, there are heterogeneous types. First we fix the type of each memory, then do list scheduling. After that, we assign other type to each memory and repeat again until we exhaust all type assignment. Then find the type assignment with minimum energy consumption. Method 3 is our TAMRS algorithm. We compare the results of our TAMRS algorithm (Method 3) with those from Method 1 and Method 2. In the list scheduling, the priority of a node is set as the longest path from this node to a leaf node [59]. The experiments are performed on a Dell PC with a P4 2.1 G processor and 512 MB memory running Red Hat Linux 9.0. In the experiments, the running time of TAMRS on each benchmark is less than one minute.

The experimental results for the three methods are shown in Table 7.2 to Table 7.4. In Table 7.4, the number of ALU is 5 and there are 5 memory types and 4 memory banks. Column "Bench." represents the name of each benchmark. Column "N" represents the number of nodes of each filter benchmark. Column "E" represents the minimum total energy consumption obtained from three different scheduling algorithms: Method 1 (Field "Method 1"), Method 2 (Field "Method 2"), and our TAMRS algorithm (Field "TAMRS"). Columns "% MD1" and "% MD2" under "TAMRS" represents the percentage of reduc-

Table 7.2. The comparison of total energy consumption with Method 1, Method 2, and TAMRS while satisfying timing constraint $L = 200$ for various benchmarks.

Memory (3 types, 2 banks), ALU (3 types), L = 200						
Benchmarks	N	Method 1	Method 2	TAMRS		
		$E(\mu J)$	$E(\mu J)$	$E(\mu J)$	% MD1	% MD2
WDF(1)	4	218	189	157	28.0	16.9
WDF(2)	12	672	564	485	27.8	14.1
IIR	16	893	751	638	28.6	15.0
DPCM	16	906	760	643	29.0	15.4
2D(1)	34	1875	1607	1359	27.5	15.4
2D(2)	4	231	204	165	28.6	19.2
MDFG1	8	492	404	349	29.1	13.7
MDFG2	8	546	450	382	30.0	15.1
Floyd	16	925	788	661	28.5	16.1
All-pole	29	1598	1356	1152	27.9	15.0
Average Reduction (%)					28.5%	15.6%

tion in total energy consumption, compared to Method 1 and Method 2, respectively. The average reduction is shown in the last row of the table.

The results show that our algorithm TAMRS can significantly improve the performance of multi-processor DSP. TAMRS shows an average 21.2% reduction in total energy consumption while satisfying timing constraint 400 using 5 types of ALU and memory. It is worthwhile to pointing out that we obtain this improvement ratio without increase the code size of applications. The reasons why our algorithm is better than Method 2 are as follows. First, Method 2 does not consider variable partition under each fixed type assignment. Second, the minimum resource scheduling and configuration algorithm has been used to improve the final performance.

In conclusion, our algorithm has several main pros: First, we study the combined effects of energy-saving and performance of memory and ALU in a systematic approach. Second, we exploit the energy saving for memory and ALU by loop scheduling and het-

Table 7.3. The comparison of total energy consumption with Method 1, Method 2 and TAMRS while satisfying timing constraint $L = 300$ for various benchmarks.

Memory(4 types, 3 banks), ALU (4 types), L = 300						
Benchmarks	N	Method 1	Method 2	TAMRS		
		$E(\mu J)$	$E(\mu J)$	$E(\mu J)$	% MD1	% MD2
WDF(1)	4	290	248	205	29.3	17.2
WDF(2)	12	882	739	610	30.8	17.5
IIR	16	1148	983	789	31.4	20.0
DPCM	16	1228	995	823	33.0	17.3
2D(1)	34	2463	2104	1687	31.5	19.8
2D(2)	4	338	267	229	32.2	14.3
MDFG1	8	612	530	421	31.2	20.5
MDFG2	8	687	589	462	32.8	21.6
Floyd	16	1229	1031	840	31.7	18.5
All-pole	29	2021	1775	1422	28.6	18.8
Average Reduction (%)					31.2%	18.5%

erogeneous type assignment. Third, data locality has been improved by using variable partition.

7.6 Related Work

Heterogeneous multi-bank memory:

To improve the overall performance, many DSPs employ Harvard architecture, which provides simultaneous accesses to separate on-chip memory banks for instructions and data [2, 35, 60, 90]. Some DSP processors are further equipped with multi-bank memory that are accessible in parallel, such as Analog Device ADSP2100, Motorola DSP56000, NEC uPd77016, and Gepard Core DSPs [35,51,60,97]. Harvesting the benefits provided by the multi-bank memory architecture hinges on sufficient compiler support. Parallel operations afforded by multi-bank memory give rise to the problem of how to maximally utilize the instruction level parallelism. To the best of our knowledge, we are the first one to con-

Table 7.4. The comparison of total energy consumption with Method 1, Method 2 and TAMRS. while satisfying timing constraint $L = 400$ for various benchmarks.

Memory (5 types, 4 banks), ALU (5 types), L = 400						
Benchmarks	N	Method 1	Method 2	TAMRS		
		$E(\mu J)$	$E(\mu J)$	$E(\mu J)$	% MD1	% MD2
WDF(1)	4	382	326	255	33.2	21.7
WDF(2)	12	1162	972	769	33.8	20.9
IIR	16	1507	1293	1001	33.6	22.6
DPCM	16	1587	1309	1056	33.5	19.3
2D(1)	34	3273	2767	2194	33.0	20.7
2D(2)	4	401	352	271	32.4	22.9
MDFG1	8	802	697	553	31.0	20.6
MDFG2	8	936	775	617	34.1	20.4
Floyd	16	1626	1356	1075	33.9	20.7
All-pole	29	2702	2335	1823	32.5	21.9
Average Reduction (%)					33.1%	21.2%

sider the combined effect of both energy and performance with heterogeneous multi-bank memory model.

Variable Partition:

Some previous work on the variable partitioning problem introduced in [51,78] tries to solve the variable partitioning problem on dual memory banks by using an interference graph. For most benchmarks used in our experiments, the variable partitioning results based on the interference graph model gives a longer schedule length. One of the limitations of the interference graph model is that it can only be applied to a directed acyclic graph (DAG), where parallelism across the loop body from different iterations is not explored. The second problem with the interference graph is that it does not incorporate sufficient information for a schedule to exploit the potential parallel memory accesses. Other dual-bank variable partitioning techniques in previous work are restricted to some specific architecture such as Motorola DSP processors [88].

Previous related work on operation parallelism can be roughly divided into two main categories: those that use compacted intermediate code as the starting point [21, 54, 78, 88]. and those that start with uncompacted intermediate code [51, 112]. Compacted intermediate code refers to the intermediate code that is compacted or scheduled by some heuristics such as list scheduling, to increase the instruction level parallelism without considering the data dependency. Since scheduling is done prior to exploring memory bank assignments, it is obvious that some memory-operation-pair combinations may be left out of consideration no matter which heuristic is used to compact the code. Thus, the approaches in the first category often fail to exploit many optimization opportunities. Techniques in the second category overcome this problem by using the uncompacted code to explore all possible pairs of memory operations as long as there are no dependencies between them. Therefore, to explore the heterogeneous multi-bank memory, we adopt the same philosophy as that of these techniques, that is, starting with the uncompacted code.

Given a program represented by a data flow graph (DFG), an undirected graph can be constructed to model the relationship among the variables in the program. The nodes in the graph represent all the local variables stored in memory. Partitioning the nodes in the graph into different groups then leads to partitioning the corresponding variables to different memory banks.

The effectiveness of such an approach relies on modeling edge weights properly to capture all relevant information. A straightforward way of assigning edge weights is to connect two nodes with an edge of weight 1 if the two corresponding variables do not have data dependencies and the memory operations involving the variables can potentially overlap [51]. The reason is that accessing such two variables in parallel may decrease the schedule length. However, such potential parallelism may not be always realizable due to certain timing constraints on the associated memory operations.

Zhuge et al. [112] introduced the concept of possibility weight to capture the likelihood of parallelizing pairs of instructions. The model does improve on the simple graph model above, but it still has some deficiencies. One deficiency of the possibility weight model is associated with simple summation of the possibility weights mentioned earlier.

Another problem with the possibility weight model is that it does not distinguish mobility overlaps within a single mobility range from those in different mobility ranges. Wang et al. [97] exploit serialism in instruction execution to trade off performance for energy savings. They use two lists to describe the edge weight in the graph model. By introducing one more dimension to the graph edge weight, they can capture the serialism information among operations and overcome the deficiencies of previous models.

Heterogeneous scheduling and assignment:

Heterogeneous assignment of special purpose architectures for real-time DSP applications has become a common and critical step in the design flow in order to satisfy the requirements of stringent timing constraint. [15, 16, 43, 96]. DSP applications need special high-speed functional units (FUs) like adders and multipliers to perform addition and multiplication operations [80]. Energy-saving task scheduling in multi-FU DSP systems has been mostly on homogeneous multiprocessors [3, 19, 110], few results considered heterogeneous systems in energy-saving real-time task scheduling [55, 80, 107]. Among the work for heterogeneous multi-FU DSP systems, Yu and Prasanna [107] considered the minimization of energy consumption for systems. The proposed algorithm is based on the *Integer Linear Programming* (ILP) without guarantees on the final solution. Luo and Jha [55] proposed list scheduling based heuristics for the scheduling of real-time tasks in heterogeneous distributed systems. However, little existing work for energy-saving scheduling in heterogeneous multi-bank memory provides guarantees on the energy consumption.

7.7 Conclusion

In this chapter, we studied the scheduling and assignment problem that minimizes the total energy without sacrificing performance on heterogeneous multi-bank memory and multi-type ALU. We proposed a highly efficient algorithm, TAMRS (*Type Assignment and Minimum Resource Scheduling*), to minimize energy consumption. TAMRS achieved a significant energy-saving by two reasons: First, use our novel type assignment algorithm with the consideration of variable partition. Second, use minimum resource sheduling to further reduce the number of resource required. A wide range of benchmarks have been tested on the experiments and the experimental results showed that our algorithm significantly improved both the energy-saving and performance for applications on heterogeneous multi-bank memory.

CHAPTER 8

CONCLUSION AND FUTURE WORK

Embedded systems are driving an information revolution with their pervasion in our everyday lives. The increasingly ubiquitous embedded systems pose a host of technical challenges different from those faced by general-purpose computers because they are more application specific and more constrained in terms of timing, power, area, memory and other resources. It becomes an important research problem to design power-aware high-performance embedded systems with various constraints and limited resources. In our research, we have attacked this problem from various aspects including high-level architecture synthesis and low power assignment. Our research focuses on understanding fundamental properties and developing models, methodologies, and algorithms for power-aware high-performance embedded systems. A lot of promising results on high-level architecture synthesis and low power optimization, have been yielded, and these results tremendously improve the state-of-the-art techniques.

Our contributions:

1. We proposed a theoretical foundation for an important problem in high-level architecture synthesis for soft real-time DSP using heterogeneous functional units (FUs), *heterogeneous assignment with probability* (HAP) problem. We modeled the execution time of each node as a random variable. We used several efficient algorithms to assign a proper FU type to each operation of a DSP application in such a way that all requirements can be met and the total cost can be minimized.

2. The solutions to the HAP problem are useful for both hard real-time and soft real-time systems. Optimal algorithms were proposed to find the optimal solutions for the HAP problem when the input is a tree or a simple path. Two other algorithms, one is optimal and the other is near-optimal heuristic, were proposed to solve the general problem. The experiments demonstrated the effectiveness of our algorithms.

3. We proposed a novel algorithm which combine *Dynamic Voltage Scaling* (DVS) and soft real-time to reduce energy consumption of uniprocessor and multiprocessor by solving *Voltage Assignment with Probability* (VAP) problem. VAP problem involves finding a voltage level to be used for each node of an date flow graph (DFG) in uniprocessor and multiprocessor DSP systems.

4. We proposed two optimal algorithms, one for uniprocessor and one for multiprocessor DSP systems, to minimize the expected total energy consumption while satisfying the timing constraint with a guaranteed confidence probability. The experimental results showed that our approach achieved significant energy saving than previous work. For example, our algorithm for multiprocessor achieved an average improvement of 56.1% on total energy-saving with 0.80 probability satisfying timing constraint.

5. We studied the energy saving issue in heterogeneous sensor networks. As we know, energy and timing are critical issues for wireless sensor networks since most sensors are equipped with non-rechargeable batteries that have limited lifetime. However, sensor nodes usually work under dynamic changing and hard-to-predict environments. We proposed a novel *adaptive online energy-saving* (AOES) algorithm to save total energy consumption for heterogeneous sensor networks.

6. Due to the uncertainties in execution time of some tasks and multiple working mode of each node, In AOES algorithm, we modeled each varied execution time as a proba-

bilistic random variable, and saved energy by selecting the best mode assignment for each node, which is called MAP (Mode Assignment with Probability) problem. We proposed an optimal sub-algorithm *MAP_Opt* to minimize the total energy consumption while satisfying the timing constraint with a guaranteed confidence probability.

7. We used loop scheduling to further extend our work of VAP problem. Low energy consumption is an important problem in real-time embedded systems and loop is the most energy consuming part in most cases. Due to the uncertainties in execution time of some tasks, we modeled each varied execution time as a probabilistic random variable. We used rotation scheduling and DVS (Dynamic Voltage Scaling) to minimize the expected total energy consumption while satisfying the timing constraint with a guaranteed confidence probability.

8. By using loop scheduling, our approach can handle loops efficiently. In addition, it is suitable to both soft and hard real-time systems. And even for hard real-time, we had good results. The experimental results showed that our approach achieves significant energy saving than list scheduling and ILP (Integer Linear Programming) voltage assignment.

9. We combined data mining and prefetching to reduce energy consumptions. The basic steps are as follows: First, we used data mining to predict the distribution of execution time and find the association rules between execution time and different inputs from history table. Then we used rotation scheduling to obtain the best assignment for total cost minimization. Finally, we used prefetching to prepare data in advance at run time. Experiments demonstrate the effectiveness of our algorithm. Our approach can handle loops efficiently. In addition, it is suitable to both soft and hard real-time systems.

10. We addressed a critical problem in multi bank on chip memory. In many high-performance DSP processors, multibank onchip memory was employed to improve performance and energy consumption. This architectural feature supports higher memory bandwidth by allowing multiple data memory accesses to be executed in parallel. However, making effective use of multi-bank memory remained difficult, considering the combined effect of performance and energy requirement.

11. We studied the scheduling and assignment problem that minimizes the total energy while satisfying performance requirements. An algorithm, TAMRS (*Type Assignment and Minimum Resource Scheduling*), was proposed. The algorithm attempts to maximum energy-saving while satisfying timing constraints. The experimental results showed that the average improvement on energy-saving is significant by using TAMRS.

Our research had produced a lot of promising results on high-level architecture synthesis and low power design. In the future, we will continue on our research in developing models, methodologies, and algorithms for high-performance, low power and secure embedded systems. One of our goals is to design an integrated embedded system development platform by extending our research results and incorporating them into this platform.

Future Research Plan:

In the future, I will keep developing models, methodologies, and algorithms for high-performance, low power and secure embedded systems by incorporating the newest techniques. Some topics are highlighted as follows.

1. *Low Power Optimization on Architecture Level.* Low power will continue to be a critical design issue and performance metric in design of embedded systems . An

architecture-level power simulator is needed for verifying power optimization techniques. There is little work in this field. I will focus on building such a power simulator. Then based on this simulator, I will develop more system-level power optimization techniques such as reducing leakage power by compiler and designing real-time operating systems incorporating dynamic voltage scheduling. Also, multi-core architecture has been adopted by almost every processor company, such as Intel and AMD. A multi-core embedded system creates enormous challenges for both hardware and software designs. For example, from software level, more advanced partition and scheduling methods are needed to minimize communication cost among cores; from hardware level, we need to consider how to design a dedicated function unit so it can be best used in a multi-core environment.

2. *Data Mining and Profiling Techniques for Embedded Systems.* Data mining is an important research area in computer science. But there is little work about using data mining for computer architecture. I will continue using data mining to save energy and improve performance for embedded systems and building a data mining engine. The building of data mining engine has several major steps: First, a history table of execution time needs to be built up. For each iteration of a loop, the engine will store the execution time of each node into the table. From history table, data-cleaning will be implemented to remove noise and inconsistent data. Then data integration and transformation will be implemented. Next, data mining is used to extract data pattern. Finally, the probability distribution function of each execution time and the association rules between inputs and execution time selection are obtained.

3. *Energy saving and performance improvement for multi-bank memory.* Many high-performance DSP processors employ multi-bank on-chip memory to improve performance and energy consumption. However, little work has been done on the combined

effect of performance improvement and energy saving. I will continue to work on the research to minimize the total energy while satisfying performance requirements.

4. *Secure Embedded Computing Framework.* Today, computer users are harassed by numerous security problems such as malicious attacks, spam, and spyware. Consumers need a simple plug-in device that can perform all protections with automatic security update. To implement this kind of device, a special purpose embedded system is needed through hardware/software codesign with a secure embedded computing framework. In this framework, many issues need to be considered. For example, an automatic backup system needs to be developed so it can automatically perform a system recovery when an instruction has been detected. Incorporating all functions such as instruction detection/prevention, spam filter, and spyware prevention into this framework with hardware/software codesign is a challenging and interesting problem.

REFERENCES

[1] I. F. Akyildiz, Y. Sankarasubramaniam W. Su, and E. Cayirci. A survey on sensor networks. *IEEE Communications Magazine*, 40(8):102–116, Aug. 2002.

[2] Analog Devices, Inc. *ADSP-21 000 Family Application Handbook Volume 1*, Norwood, MA, 1994.

[3] H. Aydin, R. Melhem, D. Mosse, and P. Alvarez. Dynamic and aggressive scheduling techniques for power aware real-time systems. In *RTSS*, 2001.

[4] C. Banino, O. Beaumont, L. Carter, J. Ferrante, A. Legrand, and Y. Robert. Scheduling strategies for master-slave tasking on heterogeneous processor platforms. *IEEE Trans. on Parallel Distributed Systems*, 15(4):319–330, 2004.

[5] O. Beaumont, A. Legrand, L. Marchal, and Y. Robert. Scheduling strategies for mixed data and task parallelism on heterogeneous clusters. *Parallel Processing Letters*, 13(2):225–244, 2003.

[6] O. Beaumont, A. Legrand, L. Marchal, and Y. Robert. Pipelining broadcasts on heterogeneous platforms. In *International Parallel and Distributed Processing Symposium IPDPS'2004*. IEEE Computer Society Press, 2004.

[7] O. Beaumont, A. Legrand, and Y. Robert. Static scheduling strategies for heterogeneous systems. *Computing and Informatics*, 21:413–430, 2002.

[8] O. Beaumont, A. Legrand, and Y. Robert. The master-slave paradigm with heterogeneous processors. *IEEE Trans. on Parallel Distributed Systems*, 14(9):897–908, 2003.

[9] P. Berman, G. Calinescu, C.Shah, and A. Zelikovsly. Efficient energy management in sensor networks. *Ad Hoc and Sensor Networks*, 2005.

[10] R. Bettati and J. W.-S. Liu. End-to-end scheduling to meet deadlines in distributed systems. In *Proc. of the International Conf. on Distributed Computing Systems*, pages 452–459, Jun. 1992.

[11] J. Bolot and A. Vega-Garcia. Control mechanisms for packet audio in the internet. In *Proceedings of IEEE Infocom*, 1996.

[12] F. Catthoor, S. Wuytack, E. D. Greef, F. Balasa, L. Nachtergaele, and A. Vandecappelle. *Custom Memory Management Methodology – Exploration of Memory Organization for Embedded Multimedia System Design*. Kluwer Academic Publishers, June, 1998.

[13] A. Cerpa and D. Estrin. Ascent: Adaptive self configuring sensor networks topologies. In *Proceedings of IEEE INFOCOM2002, New York, NY*, June 2002.

[14] Y.-N. Chang, C.-Y. Wang, and K. K. Parhi. Loop-list scheduling for heterogeneous functional units. In *6th Great Lakes Symposium on VLSI*, pages 2–7, Mar. 1996.

[15] L.-F. Chao, A. LaPaugh, and E. H.-M. Sha. Rotation scheduling: A loop pipelining algorithm. *IEEE Trans. on Computer-Aided Design of Integrated Circuits and Systems*, 16:229–239, Mar. 1997.

[16] L.-F. Chao and E. H.-M. Sha. Static scheduling for synthesis of dsp algorithms on various models. *Journal of VLSI Signal Processing Systems*, 10:207–223, 1995.

[17] L.-F. Chao and E. H.-M. Sha. Scheduling data-flow graphs via retiming and unfolding. *IEEE Trans. on Parallel and Distributed Systems*, 8:1259–1267, Dec. 1997.

[18] I. Chatzigiannakis, A. Kinalis, and S. Nikoletseas. Power conservation schemes for energy efficient data propagation in heterogeneous wireless sensor networks. In *Proceedings of the 38th annual Symposium on Simulation*, pages 60–71, Apr. 2005.

[19] J.-J. Chen and T.-W. Kuo. Multiprocessor energy-efficient scheduling for real-time tasks with different power characteristics. In *ICPP*, 2005.

[20] Y. Chen, Z. Shao, Q. Zhuge, C. Xue, B. Xiao, and E. H.-M. Sha. Minimizing energy via loop scheduling and dvs for multi-core embedded systems. In *ICPADS'05 Volume II and PDES 2005*, pages 2 – 6, Fukuoka, Japan, 20-22 Jul. 2005.

[21] J. Cho, Y. Paek, and D. Whalley. Efficient register and memory assignment for non-orthogonal architectures via graph coloring and mst algorithms. In *ACM Joint Conference LCTES-SCOPES*, pages 130–138, Berlin, Germany, Jun. 2002.

[22] V. Delaluz, M. Kandemir, and I. Kolcu. Automatic data migration for reducing energy consumption in multi-bank memory systems. In *DAC*, pages 213–218, New Orleans, LA, USA 2002.

[23] V. Delaluz, M. Kandemir, A. Sivasubramaniam, and M. J. Irwin. Hardware and software techniques for controlling dram power modes. *IEEE Trans. on Computers*, 50(11), Nov. 2001.

[24] J. Deng, Y. S. Han, W. B. Heinzelman, and P. K. Varshney. Balanced-energy sleep scheduling scheme for high density cluster-based sensor networks. *Elsevier Computer Communications Journal, Special Issue on ASWN '04*, 2004.

[25] J. Deng, Y. S. Han, W. B. Heinzelman, and P. K. Varshney. Scheduling sleeping nodes in high density cluster based sensor networks. *ACM/Kluwer Mobile Networks and Applications (MONET) Special Issue on Energy Constraints and Lifetime Performance in Wireless Sensor Networks*, 2004.

[26] A. Dogan and F. Özgüner. Matching and scheduling algorithms for minimizing execution time and failure probability of applications in heterogeneous computing. *IEEE Trans. on Parallel and Distributed Systems*, 13:308–323, Mar. 2002.

[27] J. Elson and D. Estrin. Time synchronization for wireless sensor networks. In *Proceedings of the 15th International Parallel and Distributed Processing Symposium (IPDPS '01)*, 2001.

[28] I. Foster. *Designing and Building Parallel Program: Concepts and Tools for Parallel Software Engineering*. Addison-Wesley, 1994.

[29] C. H. Gebotys and M. Elmasry. Global optimization approach for architectural synthesis. *IEEE Trans. on Computer-Aided Design of Integrated Circuits and Systems*, 12:1266–1278, Sep. 1993.

[30] J. Han and M. Kamber. *Data mining: Concepts and Techniques*. New York: Morgan-Kaufman, 2000.

[31] T. Hastie, R. Tibshirani, and J. H. Friedman. *The elements of statistical learning : Data mining, inference, and prediction*. New York: Springer, 2001.

[32] Y. He, Z. Shao, B. Xiao, Q. Zhuge, and E. H.-M. Sha. Reliability driven task scheduling for tightly coupled heterogeneous systems. In *Proc. of IASTED International Conference on Parallel and Distributed Computing and Systems*, Nov. 2003.

[33] C.-J. Hou and K. G. Shin. Allocation of periodic task modules with precedence and deadline constraints in distributed real-time systems. In *IEEE Trans. on Computers*, volume 46, pages 1338–1356, Dec. 1997.

[34] W.-C. Hsu and J. E. Smith. A performance study of instruction cache prefetching methods. *IEEE Trans. on Computers*, 47(5):497–508, May 1998.

[35] http://asic.amsint.com/databooks/digital/gepard.htm. *GEPARD Family of Embedded Software Programmable DSP Core.*

[36] S. Hua and G. Qu. Approaching the maximum energy saving on embedded systems with multiple voltages. In *International Conference on Computer Aid Design (ICCAD)*, pages 26–29, 2003.

[37] S. Hua, G. Qu, and S. S. Bhattacharyya. Energy reduction techniques for multimedia applications with tolerance to deadline misses. In *ACM/IEEE Design Automation Conference (DAC)*, pages 131–136, 2003.

[38] S. Hua, G. Qu, and S. S. Bhattacharyya. Exploring the probabilistic design space of multimedia systems. In *IEEE International Workshop on Rapid System Prototyping*, pages 233–240, 2003.

[39] C.-T. Hwang, J.-H. Lee, and Y.-C. Hsu. A formal approach to the scheduling problem in high level synthesis. *IEEE Trans. on Computer-Aided Design of Integrated Circuits and Systems*, 10:464–475, Apr. 1991.

[40] C. Im, H. Kim, and S. Ha. Dynamic voltage scheduling technique for low-power multimedia applications using buffers. In *Proc. of ISLPED*, 2001.

[41] T. Ishihara and H. Yasuura. Voltage scheduling problem for dynamically variable voltage processor. In *ISLPED*, pages 197 –202, 1998.

[42] K. Ito, L. Lucke, and K. Parhi. Ilp-based cost-optimal dsp synthesis with module selection and data format conversion. *IEEE Trans. on VLSI Systems*, 6:582–594, Dec. 1998.

[43] K. Ito and K. Parhi. Register minimization in cost-optimal synthesis of dsp architecture. In *Proc. of the IEEE VLSI Signal Processing Workshop*, Oct. 1995.

[44] D. Joseph and D. Grunwald. Prefetching using markov predictors. *IEEE Trans. on Computers*, 48(2):121–133, 1999.

[45] A. Kalavade and P. Moghe. A tool for performance estimation of networked embedded end-systems. In *Proceedings of Design Automation Conference*, pages 257 – 262, Jun. 1998.

[46] S. Kumar, T. H. Lai, and J. Balogh. On k-coverage in a mostly sleeping sensor network. In *Proceedings of the 10th Annual International Conference on Mobile Computing and Networking (Mobicom '04)*, pages 144–158, 2004.

[47] Y. W. Law, J. Doumen L. Hoesel, and P. Havinga. Sensor networks: Energy-efficient link-layer jamming attacks against wireless sensor network mac protocols. In *Proceedings of the 3rd ACM workshop on Security of ad hoc and sensor networks SASN '05*, pages Alexandria, VA, USA, 76–88, Nov. 2005.

[48] A. R. Lebeck, X. Fan, H. Zeng, and C. S . Ellis. Power aware page allocation. In *the 9th International Conference on Architectural Support for Programming Languciges and Operating Systems*, Nov. 2000.

[49] C. E. Leiserson and J. B. Saxe. Retiming synchronous circuitry. *Algorithmica*, 6:5–35, 1991.

[50] B. P. Lester. *The Art of Parallel Programming*. Englewood Cliffs, N.J.: Prentice Hall, 1993.

[51] R. Leupers and D. Kotte. Variable partitioning for dual memory bank dsps. In *IEEE Int. Conf. Acoust., Speech, Signal Process*, volume 2, pages 1121–1124, May 2001.

[52] W. N. Li, A. Lim, P. Agarwal, and S. Sahni. On the circuit implementation problem. *IEEE Trans. on Computer-Aided Design of Integrated Circuits and Systems*, 12:1147–1156, Aug. 1993.

[53] S. Y. H. Liao. *Code generation and optimization for embedded digital signal processors*. Ph.D. dissertation, Mass. Inst. Technol., Cambridge, MA, 1996.

[54] M. Lorenz, D. Kottmann, S. Bashfrod, R. Leupers, and P. Marwedel. Optimized address assignment for dsps with simd memory accesses. In *Asia South Pacific Design Automation Conference (ASP-DAC)*, pages 415–420, Yokohama, Japan, Jan. 2001.

[55] J. Luo and N. Jha. Static and dynamic variable voltage scheduling algorithms for real-time heterogeneous distributed embedded systems. In *VLSID*, 2002.

[56] H. D. Man, F. Catthoor, G. Goossens, J. Vanhoof, J. Meerbergen, S. Note, and J. A. Huisken. Architecture-driven synthesis techniques for vlsi implementation of dsp algorithms. *Proceedings of the IEEE*, 78(2):319–335, 1990.

[57] M. C. McFarland, A. C. Parker, and R. Camposano. The high-level synthesis of digital systems. *Proceedings of the IEEE*, 78:301–318, Feb. 1990.

[58] G. D. Micheli. *Synthesis and Optimization of Digital Circuits*. New York: McGraw-Hill, 1994.

[59] G. D. Micheli. *Synthesis and Optimization of Digital Circuits*. McGraw-Hill, 1994.

[60] Motorola. *DSP56000 24-Bit Digital Signal Processor Family Manual*, Schaumberg, IL, 1996.

[61] O. Mutlu, J. Stark, C. Wilkerson, and Y. N. Patt. Runahead execution: An alternative to very large instruction windows for out-of-order processors. In *IEEE HPCA-9*, Feb. 2003.

[62] K. Parhi and D. G. Messerschmitt. Static rate-optimal scheduling of iterative data-flow programs via. optimum unfolding. *IEEE Trans. on Computers*, 40:178–195, Feb. 1991.

[63] A. Parikh, S. Kim, M. Kandemir, N. Vijaykrishnan, and M. J. Irwin. Instruction scheduling for low power. *Journal of VLSI Signal Processing*, 37:129–149, 2004.

[64] N. L. Passos, E. H.-M. Sha, and S. C. Bass. Loop pipelining for scheduling multi-dimensional systems via rotation. In *Proc. 31st Design Automation Conf.*, pages 485–490, Jun. 1994.

[65] P. G. Paulin and J. P. Knight. Force-directed scheduling for the behavioral synthesis of asic's. *IEEE Trans. on Computer-Aided Design of Integrated Circuits and Systems*, 8:661–679, Jun. 1989.

[66] M. Qiu, Z. Jia, C. Xue, Z. Shao, and E. H.-M. Sha. Loop scheduling to minimize cost with data mining and prefetching for heterogeneous dsp. In *Proc. The 18th IASTED International Conference on Parallel and Distributed Computing Systems (PDCS 2006)*, Dallas, Texas, Nov. 13-15 2006.

[67] M. Qiu, Z. Jia, C. Xue, Z. Shao, and E. H.-M. Sha. Voltage assignment with guaranteed probability satisfying timing constraint for real-time multiproceesor dsp. *Jour-*

nal of VLSI Signal Processing Systems for Signal, Image, and Video Technology (JVLSI), 2006.

[68] M. Qiu, M. Liu, C. Xue, Z. Shao, Q. Zhuge, and E. H.-M. Sha. Optimal assignment with guaranteed confidence probability for trees on heterogeneous dsp systems. In *Proceedings The 17th IASTED International Conference on Parallel and Distributed Computing Systems (PDCS 2005)*, Phoenix, Arizona, 14-16 Nov. 2005.

[69] M. Qiu, Z. Shao, C. Xue, Q. Zhuge, and E. H.-M. Sha. Heterogeneous assignment to minimize cost while satisfying hard/soft timing constraints. *Submitted to IEEE Trans. on Computer*, 2006.

[70] M. Qiu, Z. Shao, Q. Zhuge, C. Xue, M. Liu, and E. H.-M. Sha. Efficient assignment with guaranteed probability for heterogeneous parallel dsp. In *Int'l Conference on Parallel and Distributed Systems (ICPADS)*, pages 623–630, Minneapolis, MN, Jul. 2006.

[71] M. Qiu, C. Xue, Z. Shao, and E. H.-M. Sha. Energy minimization with soft real-time and dvs for uniprocessor and multiprocessor embedded systems. In *Proc. The IEEE/ACM Design, Automation and Test in Europe (DATE 2007)*, Acropolis, Nice, France, April 16-20 2007.

[72] M. Qiu, C. Xue, Q. Zhuge, Z. Shao, M. Liu, and E. H.-M. Sha. Efficient algorithm of energy minimization for heterogeneous wireless sensor network. In *IEEE 17th International Conference on Application-specific Systems, Architectures and Processors (ASAP)*, Steamboat Springs, Colorado, Sep. 11-13 2006.

[73] M. Qiu, C. Xue, Q. Zhuge, Z. Shao, M. Liu, and E. H.-M. Sha. Energy minimization with guaranteed probability satisfying timing constraint via dvs and loop scheduling. *Submitted to Journal of Microprocessors and Microsystems*, 2006.

[74] M. Qiu, C. Xue, Q. Zhuge, Z. Shao, M. Liu, and E. H.-M. Sha. Voltage assignment and loop scheduling for energy minimization while satisfying timing constraint with guaranteed probability. In *Proc. 2006 IFIP International Conference on Embedded And Ubiquitous Computing (EUC'2006), Lecture Note in Computer Science (LNCS), Springer*, Korea, Aug. 2006.

[75] M. Qiu, C. Xue, Q. Zhuge, Z. Shao, M. Liu, and E. H.-M. Sha. Voltage assignment satisfying timing constraint with confidence probability. *Submitted to IEEE Trans. on Circuits and Systems II: Analog and Digital Signal Processing*, 2006.

[76] M. Qiu, C. Xue, Q. Zhuge, Z. Shao, and E. H.-M. Sha. Efficient energy saving algorithm for heterogeneous sensor network. *Submitted to International Journal of Computers and Their Applications (IJCTA), ICSA*, 2006.

[77] K. Ramamritham, J. A. Stankovic, and P.-F. Shiah. Efficient scheduling algorithms for real-time multiprocessor systems. In *IEEE Trans. on Parallel and Distributed Systems*, volume 1, pages 184–194, Apr. 1990.

[78] M. Saghir, P. Chow, and C. Lee. Exploiting dual datamemory banks in digital signal processors. In *International Conference on Architecture Support for Programming Language and Operating Systems*, pages 234–243, 1996.

[79] H. Saputra, M. Kandemir, N. Vijaykrishnan, M. J. Irwin, J. S. Hu, C-H. Hsu, and U. Kremer. Energy-conscious compilation based on voltage scaling. In *LCTES'02*, June 2002.

[80] Z. Shao, Q. Zhuge, C. Xue, and E. H.-M. Sha. Efficient assignment and scheduling for heterogeneous dsp systems. *IEEE Trans. on Parallel and Distributed Systems*, 16:516–525, Jun. 2005.

[81] S. M. Shatz, J.-P. Wang, and M. Goto. Task allocation for maximizing reliability of distributed computer systems. *IEEE Trans. on Computers*, 41:1156–1168, Sep. 1992.

[82] D. Shin, J. Kim, and S. Lee. Low-energy intra-task voltage scheduling using static timing analysis. In *DAC*, pages 438–443, 2001.

[83] A. Sinha and A. Chandrakasan. Dynamic power management in wireless sensor networks. *IEEE Design Test Comp.*, Mar./Apr. 2001.

[84] S. Slijepcevic and M. Potkonjak. Power efficient organization of wireless sensor networks. In *IEEE ICC, Helsinki, Finland*, 2001.

[85] L. Spracklen, Y. Chou, and S. G. Abraham. Effective instruction prefetching in chip multiprocessors for modern commercial applications. In *IEEE HPCA-11*, Feb. 2005.

[86] S. Srinivasan and N. K. Jha. Safety and reliability driven task allocation in distributed systems. *IEEE Trans. on Parallel and Distributed Systems*, 10:238–251, Mar. 1999.

[87] V. Srinivasan, E. S. Davidson, G. S. Tyson, M. J. Charney, and T. R. Puzak. Branch history guided instruction prefetching. In *Proc. of the 7th Int'l Conference on High Performance Computer Architecture (HPCA)*, pages 291–300, Monterrey, Mexico, Jan. 2001.

[88] A. Sudarsanam and S. Malik. Simultaneous reference allocation in code generation for dual data memory bank asips. *ACM TODAES*, 5(2), 2000.

[89] H. Tan and I. Lu. Power efficient data gathering and aggregation in wireless sensor networks. *ACM SIGMOD Record, SPECIAL ISSUE: Special section on sensor network technology and sensor data management*, 4(3):66–71, 2003.

[90] Texas Instruments, Inc. *TMS320C6000 CPU and Instruction Set Reference Guide*, Dallas, TX, Oct. 2000.

[91] T. Tia, Z. Deng, M. Shankar, M. Storch, J. Sun, L. Wu, and J. Liu. Probabilistic performance guarantee for real-time tasks with varying computation times. In *Proceedings of Real-Time Technology and Applications Symposium*, pages 164 – 173, 1995.

[92] S. Tongsima, E. H.-M. Sha, C. Chantrapornchai, D. Surma, and N. Passos. Probabilistic loop scheduling for applications with uncertain execution time. *IEEE Trans. on Computers*, 49:65–80, Jan. 2000.

[93] J. Tse and A. J. Smith. Cpu cache prefetching: Timing evaluation of hardware implementations. *IEEE Transactions on Computers*, 47(5):509–526, 1998.

[94] V. Venkatachalam and M. Franz. Power reduction techniques for microprocessor systems. *ACM Computing Surveys (CSUR)*, 37(3):195–237, Sep. 2005.

[95] C.-Y. Wang and K. K. Parhi. High-level synthesis using concurrent transformations, scheduling, and allocation. *IEEE Trans. on Computer-Aided Design of Integrated Circuits and Systems*, 14:274–295, Mar. 1995.

[96] C.-Y. Wang and K. K. Parhi. Resource constrained loop list scheduler for dsp algorithms. *Journal of VLSI Signal Processing*, 11:75–96, Oct./Nov. 1995.

[97] Z. Wang and X. S. Hu. Energy-aware variable partitioning and instruction scheduling for multibank memory architectures. *ACM Transactions on Design Automation of Electronic Systems (TODAES)*, 10(2):369–388, Apr. 2005.

[98] Z. Wang, M. Kirkpatrick, and E. H.-M. Sha. Optimal two level partitioning and loop scheduling for hiding memory latency for dsp applications. In *37th ACM/IEEE Design Automat. Conf.*, pages 540–545, June 2000.

[99] Z. Wang, T. W. O'Neil, and E. H.-M. Sha. Minimizing average schedule length under memory constraints by optimal partitioning and prefetching. *J. VLSI Signal Process. Syst. Signal, Image, Video Technol.*, 27:215–233, Jan. 2001.

[100] S. M. Weiss and N. Indurkhya. *Predictive data mining: A practical guide.* New York: Morgan-Kaufman, 1997.

[101] M. E. Wolfe. *High Performance Compilers for Parallel Computing.* Addison-Wesley, Redwood City, California, 1996.

[102] K. Wu, Y. Gao, F. Li, and Y. Xiao. Lightweight deployment-aware scheduling for wireless sensor networks. *ACM/Kluwer Mobile Networks and Applications (MONET) Special Issue on Energy Constraints and Lifetime Performance in Wireless Sensor Networks*, 2004.

[103] S. Wuytack, F. Catthoor, G. D. Jong, and H. D. Man. Minimizing the required memory bandwidth in vlsi system realizations. *IEEE Trans. on VLSI Systems*, 7(4), Dec. 1999.

[104] C. Yang, A. Lebeck, H. Tseng, and C. Lee. Tolerating memory latency through push prefetching for pointer-intensive applications. *ACM Transactions on Architecture and Code Optimization*, pages 445–475, Dec. 2004.

[105] F. Ye, G. Zhong, J. Cheng, S. Lu, and L. Zhang. Peas: A robust energy conserving protocol for long-lived sensor networks. In *Proceedings of the 23rd International Conference on Distributed Computing Systems (ICDCS '03)*, 2003.

[106] W. Ye, N. Vijaykrishnan, M. Kandemir, and M. J. Irwin. The design and use of simple power: a cycle-accurate energy estimation tool. In *the 37th Design automation Conference*, pages 340–345, June 2000.

[107] Y. Yu and V. K. Prasnna. Power-aware resource allocation for independent tasks in heterogeneous real-time systems. In *ICPADS*, 2002.

[108] L. A. Zadeh. *Fuzzy Sets as a Basis for a Theory of Possibility*, volume 1. 1996.

[109] Y. Zhang, S. Haga, and R. Barua. Execution history guided instruction prefetching. In *Intl. Conf. on Supercomputing*, pages 199–208, 2002.

[110] Y. Zhang, X. Hu, and D. Z. Chen. Task scheduling and voltage selection for energy minimization. In *DAC*, pages 183–188, 2002.

[111] T. Zhou, X. Hu, and E. H.-M. Sha. Estimating probabilistic timing performance for real-time embedded systems. *IEEE Transactions on Very Large Scale Integration(VLSI) Systems*, 9(6):833–844, Dec. 2001.

[112] Q. Zhuge, E. H.-M. Sha, B. Xiao, and C. Chantrapornchai. Efficient variable partitioning and scheduling for dsp processors with multiple memory modules. *IEEE Trans. on Signal Processing*, 52(4):1090–1099, Apr. 2004.

[113] Q. Zhuge, B. Xiao, and E. H.-M. Sha. Code size reduction technique and implementation for software-pipelined dsp applications. *ACM Transactions on Embedded Computing Systems*, 2(4):1–24, Nov. 2003.

Wissenschaftlicher Buchverlag bietet

kostenfreie

Publikation

von

wissenschaftlichen Arbeiten

Diplomarbeiten, Magisterarbeiten, Master und Bachelor Theses
sowie Dissertationen, Habilitationen und wissenschaftliche Monographien

Sie verfügen über eine wissenschaftliche Abschlußarbeit zu aktuellen oder zeitlosen
Fragestellungen, die hohen inhaltlichen und formalen Ansprüchen genügt,
und haben **Interesse an einer honorarvergüteten Publikation**?

Dann senden Sie bitte erste Informationen über Ihre Arbeit per Email
an info@vdm-verlag.de. Unser Außenlektorat meldet sich umgehend bei Ihnen.

VDM Verlag Dr. Müller Aktiengesellschaft & Co. KG
Dudweiler Landstraße 125a
D - 66123 Saarbrücken

www.vdm-verlag.de

www.ingramcontent.com/pod-product-compliance
Lightning Source LLC
LaVergne TN
LVHW022311060326
832902LV00020B/3400